Device Physics, Modeling, Technology, and Analysis for Silicon MESFET

Iraj Sadegh Amiri • Hossein Mohammadi
Mahdiar Hosseinghadiry

Device Physics, Modeling, Technology, and Analysis for Silicon MESFET

Iraj Sadegh Amiri
Computational Optics Research Group
Advanced Institute of Materials Science
Ton Duc Thang University
Ho Chi Minh City, Vietnam

Faculty of Applied Sciences
Ton Duc Thang University
Ho Chi Minh City, Vietnam

Mahdiar Hosseinghadiry
Allseas Engineering
DELFT, Zuid-Holland, The Netherlands

Hossein Mohammadi
Pasargad Higher Education Institute
Shiraz, Iran

ISBN 978-3-030-04512-8 ISBN 978-3-030-04513-5 (eBook)
https://doi.org/10.1007/978-3-030-04513-5

Library of Congress Control Number: 2018964064

This Springer imprint is published by the registered company Springer Nature Switzerland AG
The registered company address is: Gewerbestrasse 11, 6330 Cham, Switzerland

Preface

In the recent years, silicon-on-insulator metal-semiconductor field effect transistor (SOI MESFET) has attracted extensive attention due to its better mobility degradation, scalability, isolation between source-drain and substrate, immunity to hot carrier effect (HCE), and radiation hardness. However, creative approaches are needed to overcome the problems associated with downscaling. In this research, two novel designs of SOI MESFET are presented to improve the channel mobility, threshold voltage roll-off, immunity against short channel effects (SCE), and scalability. In the first design, TMG SOI MESFET, three different metals with different work functions are used as a singular gate metal. Due to this technique, two step functions appear in the profile of potential which causes two potential drops in the channel. Accordingly, the channel is screened from the variations of drain voltage to get better carrier transport and more scalability. In addition, the location of minimum potential moves toward the source and, in turn, protection against drain-induced barrier lowering (DIBL) is found. In the second design, TG SOI MESFET, better current control is gained by extending of gate metal to the side walls of the channel. This provides higher current drive, more immunity against SCEs, and further scalability. Analytical modeling and TCAD simulation for the classical SOI MESFET and the two proposed designs have been developed. The profiles of channel potential for the devices under study show the slope of bottom potential ($\Delta\Phi/\Delta x$) to the drain side at the bias condition of $V_{DS} = 0.5$ V for TMG MESFET and TG MESFET as 0.6 V/μm and 0.4 V/μm, respectively, which are much smaller than for the classical MESFET with a slope of 1.3 V/μm. It implies that the proposed devices can effectively reduce the electric field near the drain side and related short channel effects. Moreover, from the profile of threshold voltage, the threshold voltage roll-off in the classical device does not differ noticeably from that the proposed designs for longer channel gates. However, for classical SOI MESFET, the threshold voltage roll-off becomes considerable for channel lengths less than 250 nm. These similar values for TMG and TG MESFETs are 130 nm and 50 nm, respectively. It indicates the further scalability of the proposed designs. Finally, for the gate length of 100 nm, DIBL from the value of 300 mV/V, in the classical device, reduces to the values of 70 and 30 mV/V for TMG and TG MESFET, respectively. This indicates that using the proposed designs of SOI MESFET, better control of DIBL is obtained.

Ho Chi Minh City, Vietnam — Iraj Sadegh Amiri
Shiraz, Iran — Hossein Mohammadi
Delft, The Netherlands — Mahdiar Hosseinghadiry

Contents

Chapter 1
Invention and Evaluation of Transistors and Integrated Circuits

1.1 Research Background

1.1.1 The History of Transistor

In the early years of the emergence of the electronics industry, electron tubes, or vacuum tubes, were the basis of almost all electronic devices. Electron tube is an electronic device that controls transmission of electrons between two metal electrodes through the vacuum or gas within a gastight container made of glass or metal. Electron tubes mostly rely on thermionic emission of electrons provided by a hot filament or a cathode heated by the filament. The primary application of this device is to amplify weak currents or operate as a one-way rectifier for electric current [1].

In 1925, the idea of controlling the current flow by a perpendicular electric field, what today we know as metal–semiconductor FET (MESFET), was planned for the first time by Lilienfeld in 1925. He recorded a patent on the invention of the transistor in 1930 as a replacement for the electron tube, but he did not publish any paper regarding this invention [2]. However, due to the technical problems, the MEFET could not be successfully fabricated at that time and was ignored by the industry for many years [3]. Nine years after Lilienfeld, Oskar Heil followed the Lilienfeld's research and recorded another patent for the field effect transistor [4]. Afterwards, it has been demonstrated that Lilienfeld's transistor gave better results and gain [5].

In 1938, Schottky made a theory that described the rectifying behavior of a metal–semiconductor contact as dependent on a barrier layer at the surface of contact between the two materials [6]. The metal–semiconductor diodes later made on the basis of this theory are called Schottky barrier diodes.

In 1947, a research team at the Bell Laboratories including John Bardeen, Walter Brattain, and William Shockley observed that electric power gain is obtained when an arrangement of springs and wires are connected to a small crystal of germanium [7]. Using this observation, they made the first bipolar junction transistor (BJT). This transistor was constituted from two gold wires which were very closely

I. S. Amiri et al., *Device Physics, Modeling, Technology, and Analysis for Silicon MESFET*, https://doi.org/10.1007/978-3-030-04513-5_1

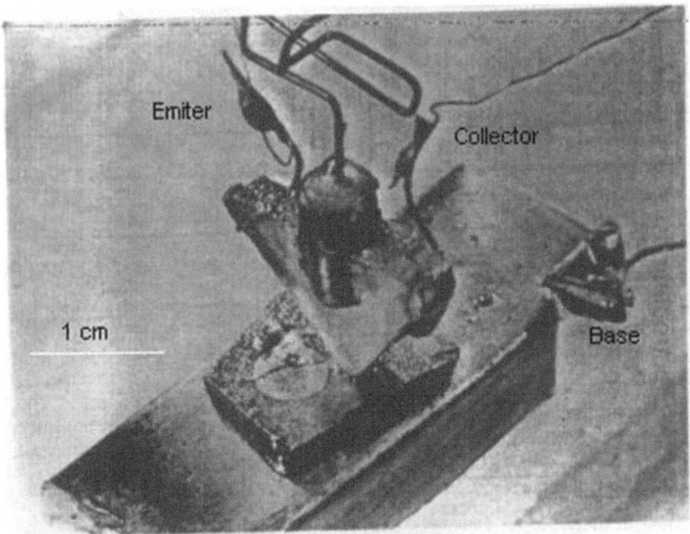

Fig. 1.1 Replica of the first working transistor invented in 1947 by John Bardeen, Walter Brattain, and William Shockley

connected to the top of germanium slice (as emitter and collector), and the third connection (base) was made to the bottom of the semiconductor [8]. This transistor could amplify an input power up to 40 times (Fig. 1.1).

In 1960, Atalla and Kahng made the first working silicon insulated-gate field effect transistor, today known as MOSFET, which had been long expected by Lilienfeld [9]. The device is principally constructed by a stack of three layers including silicon semiconductor as a base material, thermally grown native oxide as an insulator, and the metallic gate electrode as a controlling terminal. They used silicon crystals because they found that oxidation process of silicon is more comfortable than germanium. Since then, silicon had become the dominant semiconductor [10].

In 1966, Mead proposed the fabrication of metal–semiconductor field effect transistor (MESFET) by using the Schottky theory which explained the rectification between semiconductor and metal junction [11]. Eventually, Hooper and Lehrer fabricated the first prototype of MESFET using an *n*-type gallium arsenide (GaAs) epitaxial layer on semi-insulating GaAs substrate [12]. Figure 1.2 summarizes the historical growth of the electronic devices from the introduction of vacuum tubes to the beginning of the nanoscale era.

1.1.2 Integrated Circuits

The invention of solid-state devices, which replaced the electron tubes, was the beginning of microelectronics industry. Since then, the industry has experienced

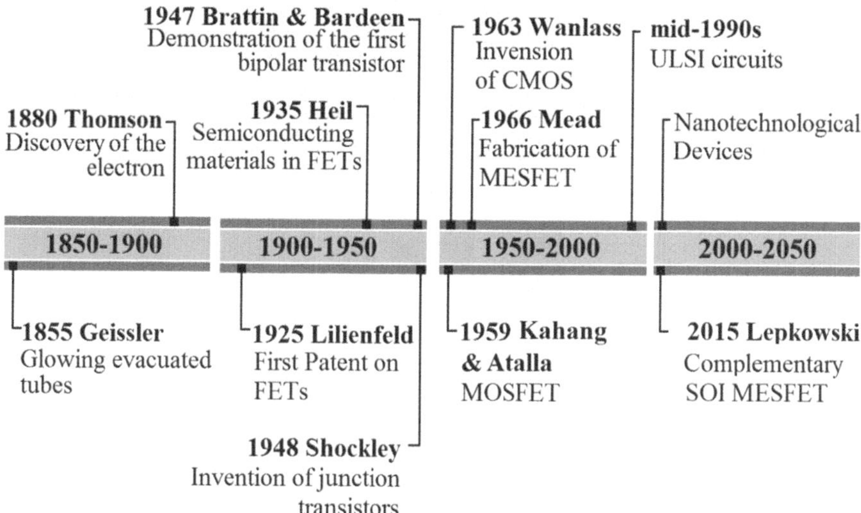

Fig. 1.2 Historic transistor timeline from vacuum tubes to the beginning of nanoscale era. Source: Morton and Gabriel 2007

phenomenal growth. In 1958, Jack Kilby[1] proposed to make transistors and other electronic components at the same time on a single semiconductor chip instead of making them separately. This idea allowed the electronic components to be fabricated by the same procedure with the same materials. In this process, the resistors could be made from bulk semiconductor, and capacitors could be made from the *p–n* junction. Kilby used a wafer of germanium and could build a flip-flop with two transistors. This was the birth of integrated circuits [13]. Six months later than Kilby, Robert Noyce[2] came up with his idea for integrated circuits and solved several practical difficulties related to Kilby's circuit including the problem of interconnection between the components on the chip. This made the integrated circuits more admissible for the mass production. The invention of complementary MOS (CMOS) by Wanlass and Sah has caused the integrated circuits (IC's) components increase to hundreds of millions [14]. Since then, integrated circuits have gradually become more complex, and the number of the components in a chip has begun to increase (Fig. 1.3).

In 1965, Gordon Moore[3] described the progress of integrated circuits in a prediction known as Moore's law. Based on Moore's law, the number of transistors per square inch of an integrated chip would be double every 2 years [15].

---

[1]Researcher of Texas Instruments.

[2]Researcher of Fairchild Semiconductor company.

[3]Co-founder of Intel.

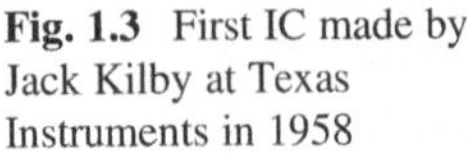

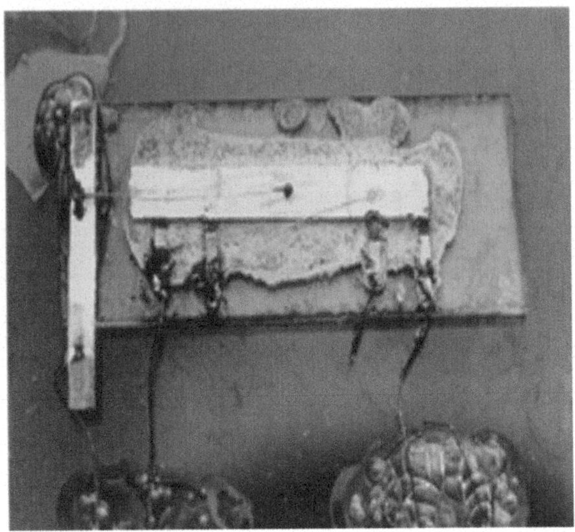

Fig. 1.3 First IC made by Jack Kilby at Texas Instruments in 1958

The transistor geometry must be shrunk every 18–24 months with respect to Moore's law. This requires robust process technology and massive research and development spending on semiconductor foundries. Over the past 50 years, due to the rapid advances in photolithography techniques, tools and pattern transfer processes, and equipment, VLSI technology has primarily followed the Moore's law. Figure 1.4 shows the level of integration over time for microprocessors. The curve shows the transistor count doubling every 2 years, per Moore's law.

Fabrication of integrated circuits with more complexity, higher speed, and a lower price has been followed on over the last decades to fulfill incessant demands for enhanced device performance. In the primary 1980s, dimension of the smallest feature size in IC exceeded the submicron, and in 2015 Intel announced its microprocessor that used transistors with 14-nm technology. Today, a modern microprocessor contains some billion transistors and a 256-gigabyte secure digital (SD) memory card, with weight of less than a gram, comprises 1012 transistors, supposing 2 bits stored in a transistor [16].

1.1.3 International Technology Roadmap for Semiconductors

Scaling is denoted as the decline of device dimensions along with several parameters by a given factor to achieve more electronic components per chip, lower power consumption, and higher speed. For a long time, the scaling trend was only based on

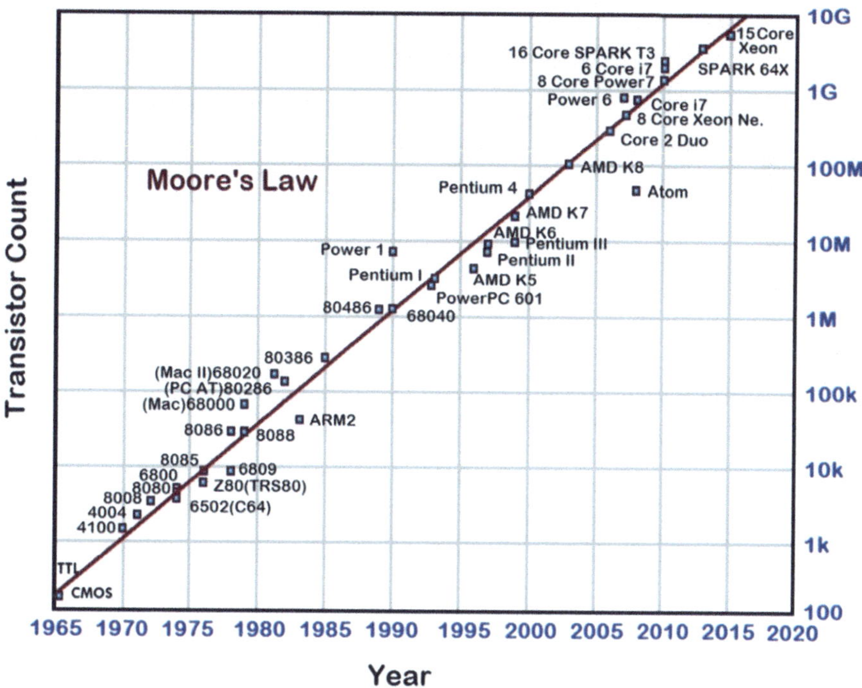

Fig. 1.4 Transistor count trend for microprocessors

lowering in feature size. The big challenge was that when the size of the transistors was about ten nanometers, due to physical limitations, the conventional scaling of MOS transistors reaches to the end of its roadmap, and the Moore' law comes to a standstill.

In the late 1980s, a group of semiconductor experts have teamed up to prepare a set of documents to evaluate the technological challenges and roadblocks related to the continuation of Moore's law and indicate the possible solutions. Therefore, the International Technology Roadmap for Semiconductors (ITRS) organization was established. The first report of ITRS was issued in 1992. Since that time, the ITRS has been publishing an annual report that avails as a criterion for semiconductor engineers. These reports clarify the kind of technology, structure, material, design, equipment, and metrology which have to be used to keep pace with the exponential growth of integration predicted by Moore's law.

Figure 1.5 displays the evolution of MOSFET gate length in production-stage integrated circuits and International Technology Roadmap for Semiconductors (ITRS) targets. As seen in the picture, the scaling of silicon MOSFET's entered into the nanometer territory around 2000. For the 0.13-μm technology node, the industry incorporated ~70-nm gate length transistors on average [17].

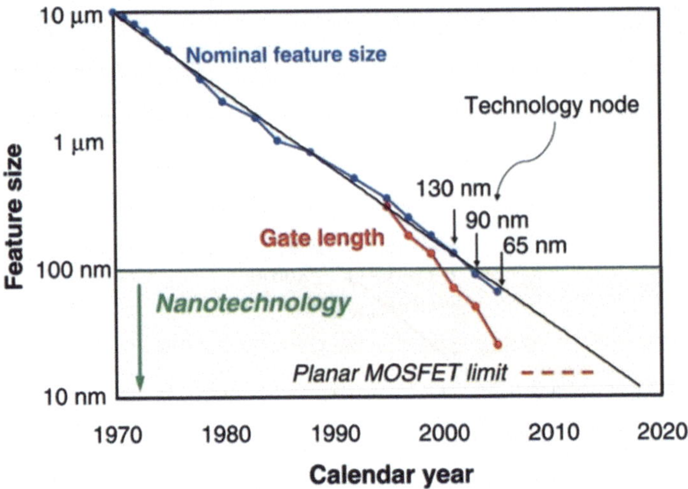

Fig. 1.5 The curve of transistor gate length versus the calendar year

1.2 Problem Statement

Nowadays, high-speed, high-performance field effect transistors (FET's) containing junction FETs (JFETs), metal–semiconductor FETs (MESFETs), metal oxide semiconductor FETs (MOSFETs), and heterostructure FETs (HFETs) are the building block of modern analog and digital integrated circuits. It is evident that the semiconductor industry owes its incredible growth to the continuous decrease in transistor dimensions. Following the trend of downscaling into the nanometer kingdom confronted with physical and thermal scaling wall. To overcome the problems associated with scaling, innovations in the device structure and utilization of new materials are needed. In this regard, various device engineering including gate engineering, source/drain engineering, isolation engineering, and channel engineering have been recommended. Also, new devices with innovative architectures with better electrostatic control over the channel such as silicon on insulator (SOI), ultra-thin body, multiple-gate and nanowire devices are developed.

Since MOSFETs are the dominant and most popular transistors used in VLSI chips, most of the innovative strategies have been done on these devices to enhance their scalability and efficiency. In comparison to the MOSFETs, MESFETs show better characteristics in terms of carrier mobility, radiation hardness, hot carrier effects, punch through effects, and scalability [18]. Moreover, MESFETs can be integrated with commercial silicon-on-insulator (SOI) or silicon-on-sapphire (SOS) CMOS ICs without changing the fabrication process or mask layers [19]. Introducing of complementary SOI MESFET, i.e., CMES gives the advantage of lower standby power consumption [20]. Also, the CMES circuitries are immune to static electricity and radiation effects, making it very ideal for applications in an ambience with a high degree of radiation. Further, the speed of CMES is in the same range as for CMOS. Moreover, due to the same construction of the main building blocks, any design using CMOS could be converted to CMES [21].

Despite the mentioned excellences of SOI MESFETs, these devices couldn't find their mark in the VLSI territory and need more serious efforts to evaluate the possibility of innovative configurations and materials on SOI MESFET devices. This leads electronic systems to increased performance with higher processing speed and lower cost. This research offers two new designs of SOI MESFETs with novelty in structure and material to achieve ICs with higher packing density together with high speed and low power dissipation suitable for military communication, satellites, aerospace, and data storage.

1.3 Objective of Research and Scope of Work

When the dimensions of conventional MOS devices are reduced to nanometers, in order to maintain the normal performance of the transistor, changes in the structure of the device and the use of new material with better quality are needed. Modeling provides the understanding of the performance and operation of semiconductor devices, and simulation validates the developed model and extremely reduces the fabrication costs and time to market. Since the advanced devices have a unique physical structure, model developers need to investigate the structure of these devices that may originally operate dissimilar to those presently used. The primary challenge facing model developers is the exact formulation and effective mathematical implementation of physically based models which can explain the device performance. This research focuses on the introduction, analysis, and modeling of novel SOI MESFET designs which eliminate roadblocks for more device scaling. The presented work contains development of new modeling methods, and also TCAD simulation. The principal objectives of this book are:

- To present a brief overview about structure and operation of MESFETs.
- To review and investigate the related works on the modeling of MESFETs.
- To present a modified analytical model for single-gate SOI MESFET.
- To propose new MESFET design, triple-material gate SOI MESFET.
- To propose new MESFET design, three-gate SOI MESFET.
- To present analytical models for the proposed devices which can explain their superiority including potential distribution, threshold voltage, subthreshold current, and drain-induced barrier lowering (DIBL).
- To use TCAD simulation to examine the accuracy of the analytical models.
- To study and discuss the performance of the innovative devices over the conventional SOI MESFETs.

1.4 Book Organization

This book is organized into five chapters. This chapter covers a brief review of the general configuration and outline of the book. First of all, the history of invention and evaluation of transistors and integrated circuits are briefly presented. After that, the concept of scaling, Moore's law, and International Technology Roadmap for Semiconductor (ITRS) are explained. Finally, the research objectives, the scope of the work, plan, and the outline of the book are expressed. Chapter 2 states the physics and fundamental concepts related to different types of field effect transistors. The necessities and various strategies related to scaling are explained. A detailed description of the origin and impact of various short-channel effects associated with downscaling and their influence on the normal operation of MOS transistors are described. The different technical solutions presented to resolve the problems caused by short-channel effects are discussed. Finally, the structures and advantages of non-classical devices and their feasibility in the settling of the short-channel effects are described. Chapter 3 focuses on analytical modeling of short-channel single-gate SOI MESFETs. In the beginning, we review the existing models presented for this device and explain the drawbacks related to each one. After that, using a new technique for solving of Poisson's equation in the conductive channel, we develop a modified two-dimensional analytical model for the device which is free of the problems associated with the previous models. By using the presented model, the subthreshold behavior of the device is displayed and discussed. Also, the impact of device parameters and bias conditions on the device performance is investigated. In Chap. 4, a new non-classical MESFET design "triple-material gate SOI MESFET" is introduced and expected to exhibit better short-channel performance. Two-dimensional analytical model of the device is derived to describe the performance of the device including surface potential, threshold voltage, and subthreshold swing. In Chap. 5, the non-classical "Three-gate SOI MESFET" is introduced and investigated by developing a three-dimensional analytical model for surface potential and threshold voltage. The model is derived by solving the 3-D Poisson's equation in the channel of the device using appropriate boundary conditions. Chapter 6 accumulates all the obtained results derived from the analytical models in Chaps. 3–5. These results take into account the various device parameters like dimensions of the channel, doping concentration, buried oxide thickness, and applied biases. For all three devices, the profile of surface potential and threshold voltage is plotted and discussed. The characteristics and excellence of the devices under study are investigated and compared with the conventional SOI MESFET. Also, the improvement in short-channel behavior of the presented devices is shown. Lastly, the TCAD simulation for each device is accomplished. The accuracy of the presented analytical models is verified by comparison with the numerical simulation results obtained by device simulator ATLAS from Silvaco. Finally, in Chap. 7, the conclusions of our research are summarized as well as some ideas for future extensions of this study are given.

References

1. S. Okamura, *History of Electron Tubes* (IOS Press, Amsterdam, 1994)
2. J.E. Lilienfeld, Method and apparatus for controlling electric currents. Google Patents, 1930
3. D. Crawley, K. Nikolic, M. Forshaw, *3D Nanoelectronic Computer Architecture and Implementation* (CRC Press, Boca Raton, 2004)
4. O. Heil, Improvements in or relating to electrical amplifiers and other control arrangements and devices. British Patent 439, 1935, pp. 10–14
5. A. Facchetti, T. Marks, *Transparent Electronics: From Synthesis to Applications* (Wiley, Chichester, 2010)
6. W. Schottky, Halbleitertheorie der sperrschicht. Naturwissenschaften **26**, 843–843 (1938)
7. J.W. Orton, *Semiconductors and the Information Revolution: Magic Crystals That Made IT Happen* (Elsevier Science, Amsterdam, 2009)
8. J. Bardeen, W.H. Brattain, The transistor, a semi-conductor triode. Phys. Rev. **74**, 230 (1947)
9. D. Kahng, M. Atalla, Silicon-silicon dioxide field induced surface devices, in *IRE Solid-State Device Research Conference*, 1960
10. G. Hellings, K. De Meyer, *High Mobility and Quantum Well Transistors* (Springer, Berlin Heidelberg, 2013)
11. C.A. Mead, Schottky barrier gate field effect transistor. Proc. IEEE **54**, 307–308 (1966)
12. W. Hooper, W. Lehrer, An epitaxial GaAs field-effect transistor. Proc. IEEE **55**, 1237–1238 (1967)
13. B. Lojek, *History of Semiconductor Engineering* (Springer, Berlin Heidelberg, 2007)
14. F. Wanlass, C. Sah, Nanowatt logic using field-effect metal-oxide semiconductor triodes, in *1963 IEEE International Solid-State Circuits Conference*. Digest of Technical Papers. IEEE, 1963, pp. 32–33
15. G. Moore, Cramming more components onto integrated circuits. Electron. Mag. **38**, 82–85 (1965)
16. J.P. Colinge, J.C. Greer, *Nanowire Transistors: Physics of Devices and Materials in One Dimension* (Cambridge University Press, Cambridge, 2016)
17. S.E. Thompson, S. Parthasarathy, Moore's law: the future of Si microelectronics. Mater. Today **9**, 20–25 (2006)
18. C.H. Suh, Analytical model for deriving the threshold voltage of a short gate SOI MESFET with vertically non-uniformly doped silicon film. IET Circ. Device Syst. **4**, 525–530 (2010)
19. W. Lepkowski, M.R. Ghajar, S.J. Wilk, N. Summers, T.J. Thornton, P.S. Fechner, Scaling SOI MESFETs to 150-nm CMOS technologies. IEEE Trans. Electron. Device **58**, 1628–1634 (2011)
20. W. Lepkowski, S.J. Wilk, T.J. Thornton, Complementary SOI MESFETs at the 45-nm CMOS node. IEEE Electron. Device Lett. **36**, 14–16 (2015)
21. K.E. Bohlin, P.A. Tove, U. Magnusson, J. Tiren, Complementary silicon MESFET technology. Electron. Lett. **23**, 205–206 (1987)

Chapter 2
General Overview of the Basic Structure and Operation of a Typical Silicon on Insulator Metal–Semiconductor Field Effect Transistor (SOI-MESFET)

2.1 Basics of SOI-MESFET

The term MESFET is taken from metal–semiconductor field effect transistor. As shown in Fig. 2.1, ohmic contacts are used for source and drain terminals, but the gate electrode is formed by a metal–semiconductor contact (Schottky contact). The fabrication technology of Schottky barriers provides fabrication of MESFETs in tiny dimensions with high accuracy. These devices are mostly made using silicon or III–V group compounds like GaAs, InP, GaN, and SiC [1]. The absence of oxide layer in the building of MESFET has caused no capacitance formed in the device.

Since the Fermi level of metal and semiconductor is different, a potential barrier is formed to balance the Fermi levels of the junction when they are connected. The Fermi level of a semiconductor is more than for the metal. Therefore, the excess n-type carriers in the semiconductor will leave the bound positive charges behind, shape the depletion region, and spillover into the metal. The potential barrier between the gate and the conducting area above the depletion region plays the role of insulator and separates the gate from the moving electrons in the conducting channel. The conducting channel is shaped by the gate depletion area and the semi-insulating substrate. Applying a potential to the drain causes electrons travel from the source to the drain. Variation of the gate voltage changes the shape of the depletion region and consequently the current flow.

Since the mobility of electrons is higher than holes, n-channel MESFETs are mainly more popular than p-channel. However, both types of MESFETs are attractive because complementary technology, which is the leading technology for fabrication of integrated circuits with the advantage of low-power dissipation, is based on a combination of both n-channel and p-channel devices [2]. For proper operation of MESFET, the appropriate biases of gate-source and gate-drain are needed. These two biases vary the gate depletion width and electric field that, in turn, control the channel current. Regarding the values of V_{GS} and V_{DS}, three operation modes of triode, saturation, and pinch-off are available.

I. S. Amiri et al., *Device Physics, Modeling, Technology, and Analysis for Silicon MESFET*, https://doi.org/10.1007/978-3-030-04513-5_2

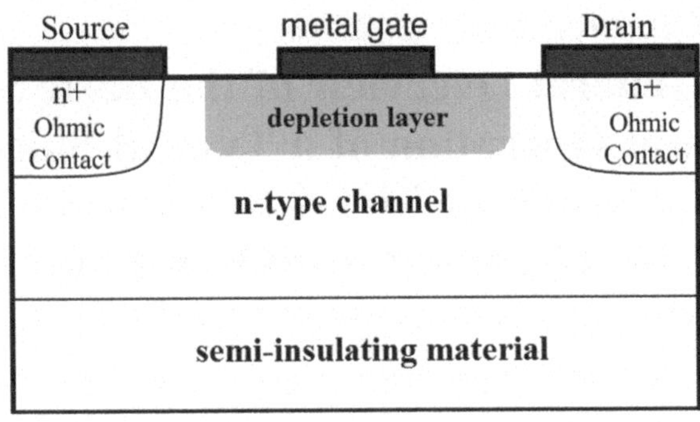

Fig. 2.1 Schematic illustration of n-channel MESFET, showing its difference with JFET in the formation of gate capacitor

For n-channel MESFET, when the gate-source voltage is zero and drain-source voltage is small, the depletion area below the Schottky gate is almost thin. In this case, a low drain current, which linearly depends on voltage, flows in the channel. Increasing of drain voltage accelerates the movement of electrons in the channel and the drain current increases. The potential drop along the channel causes the electric field below the gate, corresponding to the distance from the source, increases. In this case, the voltage between source and drain varies from zero (at the source) to V_{DS} (at the drain). This event increases the reverse bias of the Schottky barrier as one move between the source and drain. When V_{DS} increases, the width of depletion region becomes broader and, in turn, the width of conducting channel reduces. Therefore, the channel resistance increases and the current decreases. This area of device operation is known as linear or triode region (Fig. 2.2a). More increasing of drain voltage induces more reverse bias of gate and channel toward the drain. Hence, the width of depletion area becomes wider from drain to source. In a particular value of drain voltage, the depletion area reaches the semi-insulating substrate. Here, the source and drain are entirely separated or pinched-off by depletion area (Fig. 2.2b). The current after this point is saturated and the drain voltage is named saturation voltage. If drain voltage increases after the saturation, the pinch-off point moves toward the source (Fig. 2.2c). As shown in Fig. 2.2d, a further negative value of gate bias reduces the channel opening and also the drain current [3].

Figure 2.3 displayed the I–V curve of an n-type MESFET for different values of V_{GS}. As the design of MESFET is independent of the diffusion of dopant, the channel can be fabricated thin enough so that the channel thickness is blocked by Schottky barrier even without any bias applied to the gate. Therefore the MESFET can be available either as a "normally-on" or "normally-off" device [4].

In normally-off MESFET, which is known as enhancement mode, the channel is totally depleted by the gate built-in potential even at zero gate voltage. Therefore, no channel exists between the source and drain at zero gate voltage. In this case, applying a positive bias on the gate obtains the drain current in the channel. The threshold voltage of normally-off device is positive [5]. In normally-on MESFET, which is known as depletion mode, the channel naturally exists even without gate bias. Applying a negative voltage to the gate causes the depletion width of Schottky barrier extends into the semi-insulating substrate and limits the current between source and

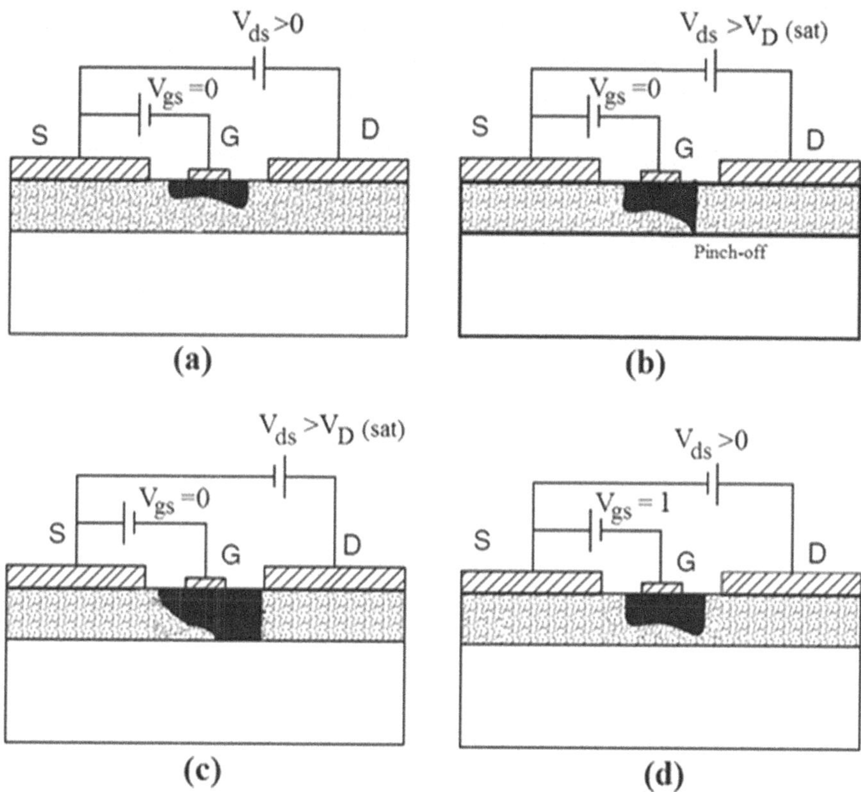

Fig. 2.2 Channel situations of MESFET at different biases (**a**) low drain voltage, (**b**) onset of saturation, (**c**) beyond the saturation, and (**d**) high negative gate voltage

drain. The certain minimum voltage needed to reach threshold at cutoff is called threshold voltage. Figure 2.4 compares the structure of normally-on and normally-off MESFETs.

2.2 CMOS Scaling

The phenomenal progress in the integrated circuits technology is primarily due to the growing demands for electronic devices with higher performance and more functionality at a reduced cost. The design of very large scale integrated (VLSI) circuits needs that the packing density of electronic components increases as much as possible. To achieve this goal, the size of electronic components must shrink. Scaling of MOS devices is defined as the systematic shrinking of total dimensions of the electronic components as the existing technology allows. Scaling is advantageous due to raising the number of dies in a wafer, to improve the speed of the circuit

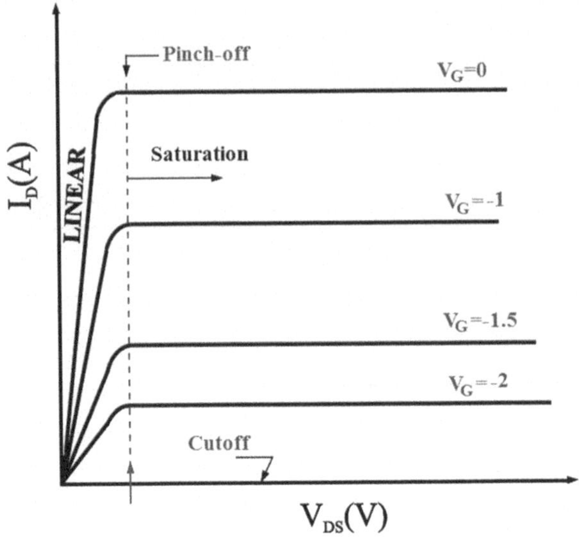

Fig. 2.3 Current–voltage characteristics of an n-type MESFET with $V_{GS} = 0$ (top curve), -1 V, -1.5 V, and -2 V

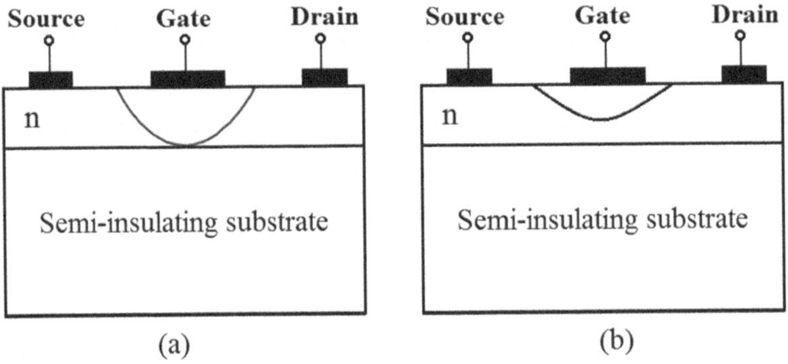

Fig. 2.4 Different modes of MESFET (**a**) normally-off, (**b**) normally-on

and the dynamic power consumption [6]. In the scaling, the geometric ratios of larger devices must be preserved. Therefore, all components of the integrated circuit must be shrunk by the same. Currently, two strategies of scaling are available: constant field scaling (full scaling) and constant voltage scaling.

2.2.1 Constant Field Scaling

In this mode of scaling, the geometry dimensions of the device are scaled down by the factor of S, but the magnitude of internal electric fields remains constant. To achieve this object, all potentials must scale down proportionally, by the same scaling factor. Likewise, the gate capacitance is also scaled due to the reduction of device geometry. But some parameters like thermal voltage and energy band gap do not be scaled. Therefore, built-in potential and surface potential, which depends on energy gap, cannot be scaled too. Accordingly, depletion width does not scale as much as other parameters which result in degraded secondary effects. Also, the threshold voltage cannot be easily scaled. However, constant field scaling offers a proper framework for device scaling with keeping the reliability of the device [7].

2.2.2 Constant Voltage Scaling

The scaling of the power supply by the same factor as the device dimension is not always feasible. Thus, constant voltage scaling is commonly desired. In this mode of scaling, all dimensions of the MOSFET are reduced by a factor of S, similar to full scaling, but the power supply voltage and all terminal voltages remain constant. To keep the charge–field relations, we need to multiply the doping densities by a factor of S2. Table 2.1 compares the scaling factors of the two scaling modes for all major dimensions, potentials, and doping densities [8].

Table 2.1 Impact of scaling on MOS device characteristics

Quantity	Before scaling	Constant field	Constant voltage
Channel length	L	$L' = L/S$	$L' = L/S$
Channel width	W	$W' = W/S$	$W' = W/S$
Gate oxide thickness	t_{ox}	$t'_{\text{ox}} = t_{\text{ox}}/S$	$t'_{\text{ox}} = t_{\text{ox}}/S$
Oxide capacitance	C_{ox}	$C'_{\text{ox}} = S \cdot C_{\text{ox}}$	$C'_{\text{ox}} = S \cdot C_{\text{ox}}$
Junction depth	x_j	$x'_j = x_j/S$	$x'_j = x_j/S$
Power supply voltage	V_{DD}	$V'_{\text{DD}} = V_{\text{DD}}/S$	V_{DD}
Threshold voltage	V_{th}	$V'_{\text{th}} = V_{\text{th}}/S$	V_{th}
Substrate doping	N_A,N_D	$N'_A(N'_D) = S \cdot N_A(N_D)$	$N'_A(N'_D) = S^2 \cdot N_A(N_D)$
Drain current	I_D	$I'_D = I_D/S$	$I'_D = S \cdot I_D$
Power dissipation	P	$P' = P/S^2$	$P' = S \cdot P$
Area	$W \cdot L$	$W'L' = WL/S^2$	$W'L' = WL/S^2$

2.3 Short-Channel Effects Associated with Device Scaling

A FET is defined as a short channel when the channel length is scaled down to become comparable to drain depth and source depth. The proximity between the source and the drain helps a better control of the channel area which can be affected by electric field lines of the source and drain. So some unwanted effects arise which disturb the normal operation of the device. These effects are famous as short-channel effects (SCEs) have several damaging effects on device performance. Some significant short-channel effects are drain-induced barrier lowering (DIBL), hot carrier effects (HCE), impact ionization, channel length modulation, surface scattering, and velocity saturation of carriers.

2.3.1 Drain-Induced Barrier Lowering and Punch-through

Drain-induced barrier lowering or DIBL is referred to the decrease of the threshold voltage at higher drain voltages. The movement of carriers in the channel is related to the condition of the inversion layer. When the gate bias is not sufficient, the mobile carriers (electrons) in the channel meet a potential barrier which stops them from flowing. Increasing of the gate voltage reduces this potential barrier and allows the carriers to flow under the impact of the channel electric field. The carrier transportation can be performed perfectly if the barrier height depends only on the gate voltage.

In short-channel MOSFETs, the potential barrier is a function of both the gate voltage (V_{GS}) and the drain voltage (V_{DS}). When a larger drain voltage is applied, the depletion region of drain encroaches the area under gate and barrier lowering of electrons in channel occurs. DIBL provides the movement of electrons between the source and the drain, even if the gate to source voltage is less than the threshold voltage [9]. The DIBL phenomenon is often accompanied by punch-though. Punch-through occurs when the depletion layers around the drain and source regions join together and form a single depletion area as shown in Fig. 2.5.

In this case, the field below the gate strongly depends on the drain to source voltage so that the gate voltage cannot control the current flow between source and

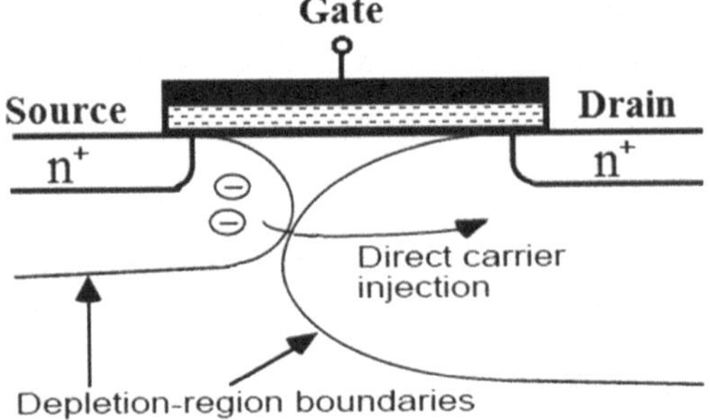

Fig. 2.5 Representation of punch-through in short-channel MOSFET

drain. Punch-through causes the current rapidly increases with increasing of drain voltage. This effect increases the output conductance and limits the maximum operating voltage of the device.

2.3.2 *Hot Carrier Effect*

One of the various short-channel effects in CMOS scaling is hot carrier effect (HCE) or hot electron effects (HEE) which referred to the device degradation caused by hot carrier injection. The term "hot carriers" is devoted to the carriers (electrons for n-channel devices) that gain sufficient kinetic energy when they accelerate under a high electric field. When the device dimension shrinks regardless of the power supply, the electric field in the device increases. This high electric field causes the carriers gain high enough kinetic energy to leave their place and tunnel in the oxide layer as shown in Fig. 2.6. These hot carriers, which are trapped in the oxide layer, will change the threshold voltage and permanently disturb the normal operation of the MOSFET.

2.3.3 *Impact Ionization*

Impact ionization is another undesirable effect related to short-channel devices. The presence of a high longitudinal field, in short-channel devices, causes electrons accelerate toward the drain area. These hot electrons can impact silicon atoms and therefore generate electron-hole pairs by ionization. This phenomenon is continued by the generation of new electron-hole pairs due to impact ionization as shown in Fig. 2.7. Most of the generated electrons travel to drain and give an increase in the drain current which, in turn, decreases the output impedance of the transistor. The generated holes travel toward the substrate and form a parasitic substrate current which can bring some undesirable effects like noise and latch-up.

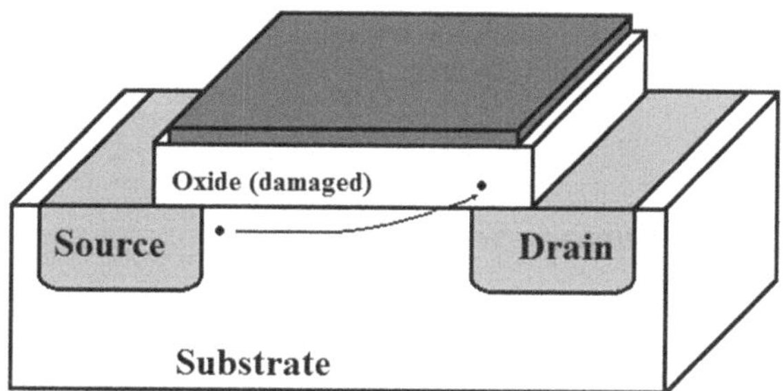

Fig. 2.6 Hot carrier manages to enter and gets trapped in the oxide layer

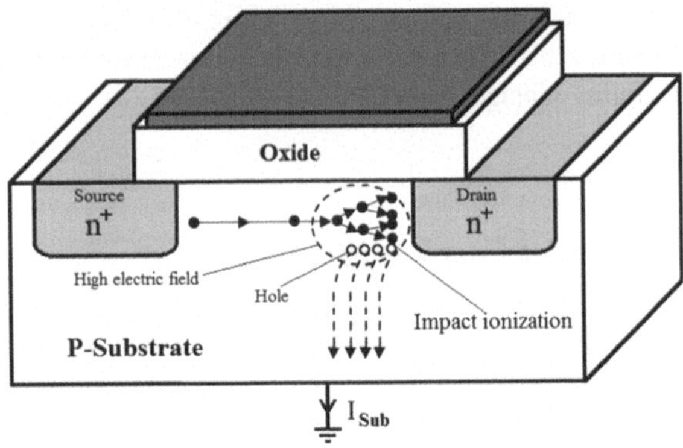

Fig. 2.7 Impact ionization in MOSFET

Moreover, further generation of electron-hole pairs can increase drain current abruptly so that the breakdown happens. The substrate current can also effect on the adjacent devices on the chip.

2.3.4 *Channel Length Modulation*

Channel length modulation (CLM) is referred to shortening of the length of the inverted channel with an increase in drain bias for high drain biases when the MOSFET is in saturation. When the channel pinches off at the drain and the drain voltage increases to exceed V_D(sat), the pinch-off point starts to move from the drain toward the source. Accordingly, the inverted channel length shortens, and the length of inversion layer continuously decreases with more increasing of drain bias. The MOSFET under this condition operates as if its channel length L was shortened by ΔL, so that drain current becomes larger than its value at the onset of saturation when ΔL was zero. Thus, in the saturation mode, the channel region splits into two different regions: one on the source side where the GCA is valid, and a second on the drain side where the GCA is violated. If the channel of the device is long enough, the length of the pinch-off area is negligible. But if the device channel is short, as shown in Fig. 2.8, the extension of the non-inverted region toward the source causes the pinch-off area occupies a considerable part of the channel length. The decline of channel length results in increases of current with drain bias and reduction of output resistance in the saturation region. The CLM effect more happens in the devices with a low-doped substrate [10].

2.3.5 *Surface Scattering*

In operation of MOS devices, the electric field of the gate causes the carriers, which are traveling along the channel, are absorbed to the surface. These carriers are limited

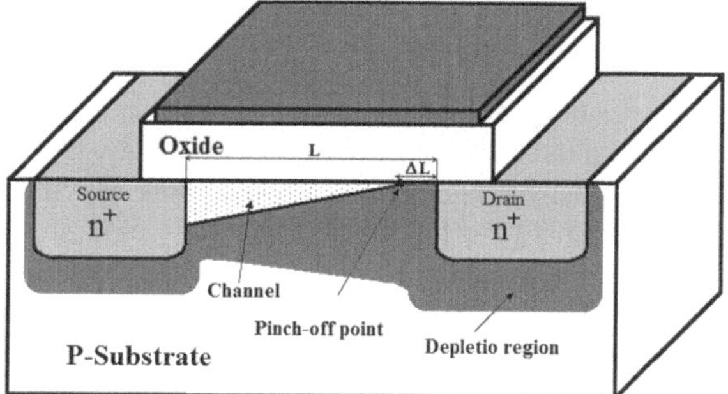

Fig. 2.8 Channel length modulation effect

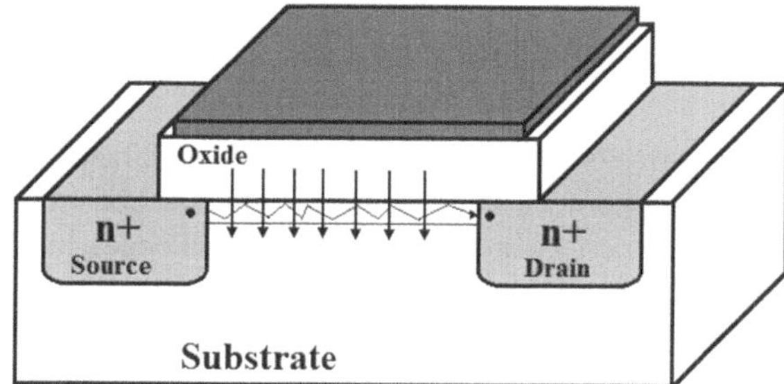

Fig. 2.9 Zigzag movement of carriers affected by gate electric field

in a small thickness near the interface of silicon oxide. Consequently, they crash and bounce against the surface during their movement. This phenomenon causes the carriers follow a zigzagging path which is depicted in Fig. 2.9.

Surface scattering efficiently degrades the surface mobility of the carriers. In short-channel devices, the lateral electric field created by drain voltage is stronger. As a result, surface scattering becomes heavier, reducing the efficient mobility in comparison with long channel devices.

2.3.6 Velocity Saturation of Carriers

This phenomenon, which is one of the significant SCEs, has a damaging influence on device performance by degrading the movement of carriers in the device. In the long channel MOSFETs, the increase of drain voltage increases the corresponding drain current. When the drain voltage reaches the pinch-off point, the drain current becomes saturated. But in short-channel devices, the electric field across the channel is very high so that the velocity of carrier becomes saturated at lower drain voltages. In this case, the carrier velocity won't rise by more increasing of the applied electric field.

2.4 Quantum Mechanical Effect [11]

Since the first integrated circuit is invented, the technology progress mostly has focused on three issues: improved functionality, higher performance, and decreased cost. For many years, the conventional bulk MOSFETs were assumed as the only available devices to follow the technology progress. During this period, the architecture and operation principle of scaled MOS devices have fundamentally not been changed. The continuous shrinking of physical dimensions has carried improved performance and lower cost.

2.5 Technical Solutions for Further Scaling of MOS Devices

In the year 2000, the downscaling trend of silicon MOSFETs encroached into the submicron territory when the gate length had been reduced to about 100 nm. Hence, the device performance faced several challenges like short-channel effects (SCEs) and the junction leakage current which make challenging to preserve the power consumption at a satisfying level. In this condition, the following of Moore's law with classical MOS devices became impossible. Therefore, unconventional devices are considered as a candidate to overcome scaling related difficulties to continue the scaling trend in the submicron region.

Regarding international technology roadmap for semiconductor (ITRS), the non-classical MOS devices are categorized in five groups of transport-enhanced FETs, UTB-SOI FETs, source/drain engineered, gate engineered, and multiple gate transistors [12]. Transport-enhanced FETs include the devices that their current drive is increased by improving the average mobility of carriers in the channel. It is achieved by using new channel material like germanium, silicon-germanium, or III–V compound semiconductors which have higher carrier mobility than silicon. The UTB SOI FETs contain a very thin fully depleted body ($t_{si} \leq 10$ nm) to provide better electrostatic control of the channel by the gate. Source and drain engineering techniques are used to maintain the resistance of the source and drain to a suitable fraction of the channel resistance (about 10%). Two methods are used to address this matter. The first is replacing of heavily doped source and drain areas by silicide to form Schottky contacts which minimize the parasitic series resistance [13]. Second is the degrading of fringing/overlap gate. Since the device dimensions are scaled down, the current drivability of device is suffered from the detrimental impact of the parasitic capacitance between the gate and source/drain areas. This event is due to the overlap between the source/drain to the gate which increases the gate overlap capacitances. In the non-overlapped gate structure, the gate electrode is fabricated so that no overlap occurs with source and drain which results diminishing of the parasitic fringing/overlap capacitance [14]. The gate engineering method such as the dual metal gate (DMG) structure uses two different materials with different work-functions merged laterally together as a single gate. Due to this configuration, the electron velocity and electric field along the channel abruptly rise near the interface of two gate materials, causing improved gate transport efficiency. Multiple gate transistors are non-classical structures which have been proposed to perform the electrostatic integrity.

2.6 Architecture-Based Non-classical Devices

In the recent years, according to technical solutions, various types of non-classical devices including silicon-on-insulator (SOI) devices, double-gate (DG) MOSFETs, triple-gate (TG) MOSFETs, quadruple-gate (QG) MOSFETs, strained MOSFETs, and dual material gate (DMG) MOSFET's have been invented.

2.6.1 Silicon-On-Insulator

Silicon-on-insulator (SOI) technology is one of the solutions to continue the scaling trend of electronic devices which has became practical in early 1990 by IBM in 1998 [15] and it was introduced at around 45 nm technology node mass production [16]. Devices with SOI substrate offer a higher performance such as better channel mobility and diminished short-channel effects over bulk devices [17]. The substrate, material, and fabrication process of SOI devices are same as bulk devices, but the fundamental difference is that the source, drain, and body of SOI devices are placed on an insulating oxide layer unlike the bulk MOSFETs that the silicon channel is directly placed on the substrate. The cross-section view of a basic n-MOS on bulk and SOI devices is shown and compared in Fig. 2.10. The SOI configuration consists of a thin layer of silicon over an insulator which is called "buried oxideor BOX."

The thickness of the silicon film is typically between 50 and 200 nm, while the buried oxide thickness usually ranges between 80 and 400 nm. The insulating layer is created by flowing oxygen onto a plain silicon wafer and then heating the wafer for oxidizing the silicon, thereby forming a uniform buried layer of silicon dioxide. The transistor is enclosed by SiO_2 on all sides.

SOI technique brings many advantages over bulk MOSFETs in aspects of the device and circuit performance. First of all, it enhances device performance by diminishing parasitic capacitances due to isolating of the junction from bulk silicon which, in turn, reduces the delays in CMOS digital circuits. In other words, using SOI technology increases the speed of digital integrated circuits. Moreover, due to the decreasing of leakage current through the buried oxide layer (BOX), the power delay of CMOS integrated circuits immensely reduces. In summary, SOI offers high-speed and low-power advantages. In the device feature, since the channel is isolated from the substrate, the SOI devices exhibit excellent latch-up immunity [18]. Furthermore, a finite number of electric field lines can propagate into the silicon film and the channel area. This event efficiently degrades the short-channel effects. Also, the

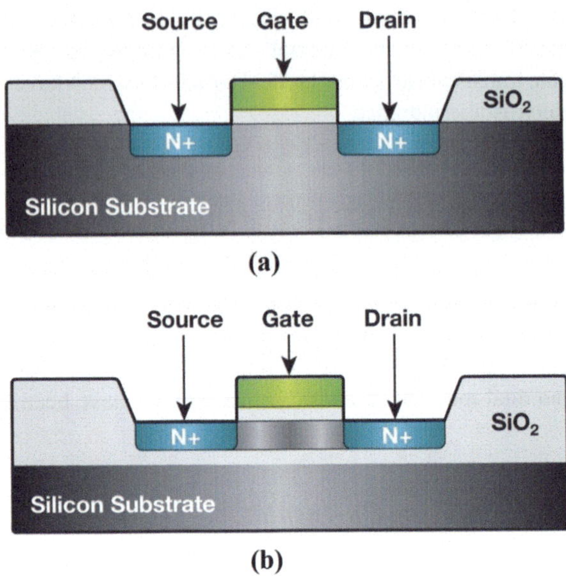

Fig. 2.10 A typical n-MOS transistor with bulk (**a**) and SOI (**b**) structure

channel mobility of SOI devices is more than their bulk counterparts [19]. Finally, the SOI-MOSFETs have higher device density than bulk MOSFETs. When the device dimensions are shrunk, doping densities must be increased to preserve device performance. Hence, for SOI devices, lower doping densities are required [20].

Regarding the thickness of the silicon layer, SOI-MOSFETs are categorized into two groups of fully depleted (FD-SOI) and partially depleted (PD-SOI). If the silicon film is thin enough that the depletion zone below the gate fills the entire depth of the body, in this manner, the device is named fully depleted. If the silicon film is thicker than the maximum gate depletion width, a part of the body cannot be depleted, and a floating body is shaped. These transistors are named partially depleted. Figure 2.12 shows the difference between two kinds of SOI structures.

Partially Depleted SOI

PD-SOI transistors are fabricated on a relatively thick layer of silicon so that the depletion depth of the channel does not occupy the whole thickness of the silicon layer. The operation and performance of a PD-SOI-MOSFETs are almost similar to bulk MOSFETs. The difference is that, due to the complete isolation of the channel from the substrate, the bottom of the silicon film of PD-SOI-MOSFET is electrically floating. As a result, a neutral body region is formed below the gate depletion boundary. The accumulation of positive charges generated by the impact ionization near the drain region in the floating body alters the potential of the body which, in turn, varies the threshold voltage of the device. This phenomenon, which is the foremost parasitic effect in SOI-MOSFETs, is called floating body effect (FBE).

FBE may create an increase, or kink, in the output characteristics of the device in high inversion which is called kink effect (Fig. 2.11). Kink effect is more observed in n-channel devices because the impact ionization rate of electrons is more than holes.

FBE also brings the history effect which is the dependence of the threshold voltage on its former states. When a PD-SOI device stays workless for a long time, the body voltage reaches to equilibrium condition by the leakage currents of the source and drain junctions. Now if the device begins to switch, the charge may spill off the body and significantly changes the body voltage and threshold voltage. In analog devices, the FBE is called kink effect. Many of the detrimental consequences of FBE could be removed by using a body contact for MOSFET, but this is not admirable.

Fully Depleted SOI

In fully depleted SOI devices, the top silicon layer is very thin as the channel depletion area extends through the whole thickness of the silicon layer before the threshold condition happens. As a result, the positive charges cannot accumulate in the body area, and consequently, the neutral region does not arise. This results in diminishing of FBE [10]. FD-SOI devices are also certainly free of kink effect. Since the entire of the silicon film is depleted, majority carriers can penetrate more freely into the source area. This avoids the accumulation of excess carriers [21] (Fig. 2.12).

Although the superiority of FD-SOI over PD-SOI, the normal operation of these devices is suffered from interface coupling effect where all device parameters are affected by the voltage of the back gate which is formed at the buried oxide. Moreover, PD-SOI has priorities regarding manufacturing, design, multiple threshold voltages, and short-channel effects. The FD-SOI devices are also called thin-film because the silicon film thickness of these devices is always less than the maximum depletion width.

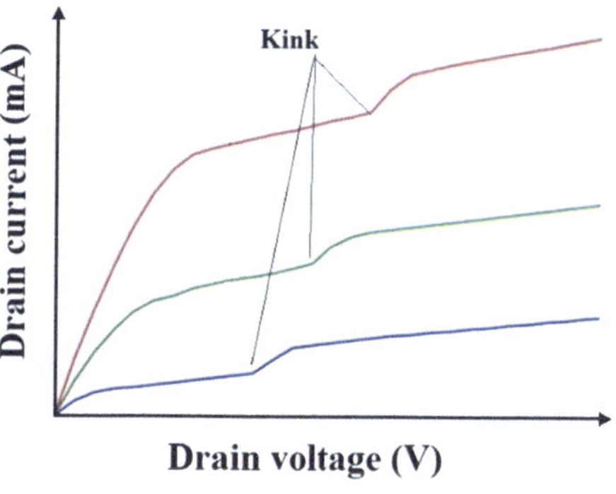

Fig. 2.11 Plot of Kink effect

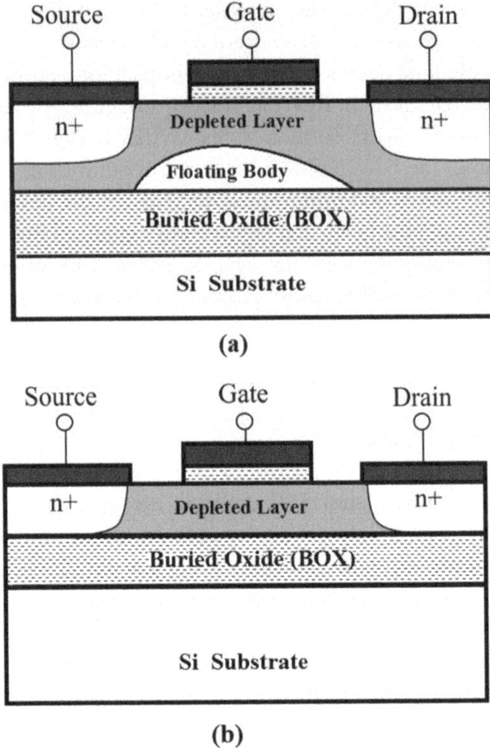

Fig. 2.12 Schematic structures of SOI-MOSFETs. (**a**) Partially depleted and (**b**) fully depleted, displaying distinction between their depleted layers

2.6.2 *Devices with Multiple Gates*

A multiple gate FET (MuGFET) or multi-gate transistor refers to a MOSFET which contains a channel wrapped by several gates on several sides, providing the shortest channel length promising for a particular gate oxide thickness. The multi-gate devices are currently considered as the next generations of transistors due to their grand electrostatic control and also compatibility with CMOS process. The multi-gate devices can be fabricated on bulk or SOI substrate. Various kinds of multi-gate devices including double-gate, triple-gate, and surrounding gate are available. It is a notable that some multi-gate devices are known by different names [22].

Double-Gate MOS Devices

Better scalability and a remarkable decrease of short-channel effects are earned by adding a second gate electrode at the other side of the body of a fully depleted SOI device. This device configuration is called double-gate (DG) and was proposed for the first time by Sekigawa and Hayashi in 1984 [23]. The key advantage of double-gate structure on SOI-MOSFET is the enhanced control of the channel by suppressing the effect of drain electric field on the channel which results in the reduction of short-channel effect. The additional gate allows the DG devices to have two channels of current flow [24]. This property provides several modes of operation for DG devices: First, the front channel is conducting while the back channel is either depleted or accumulated. Second, both channels are conducting, but one or both of the channels are in weak or strong inversion [25]. Double-gate devices are mostly found in three configurations of planar, FinFET, and MIGFET. In a planar double-gate device, a second gate is placed below the channel as illustrated in Fig. 2.13a. The FinFET, which is the most popular style of double-gate MOSFET, consists of a thin silicon body enclosed by gate electrodes which form the body of the device with a dielectric layer called hard mask on top of the silicon fin (Fig. 2.13b). The task of the hard mask is to prevent charge-sharing between two adjacent gates.

A double-gate device can be biased by independent gate potentials if the two gates are not tied together. In this case, the device is called multiple independent gate field effect transistor (MIGFET). The advantage of MIGFET is that two gates can be biased separately. The critical property of the MIGFET is that the threshold voltage

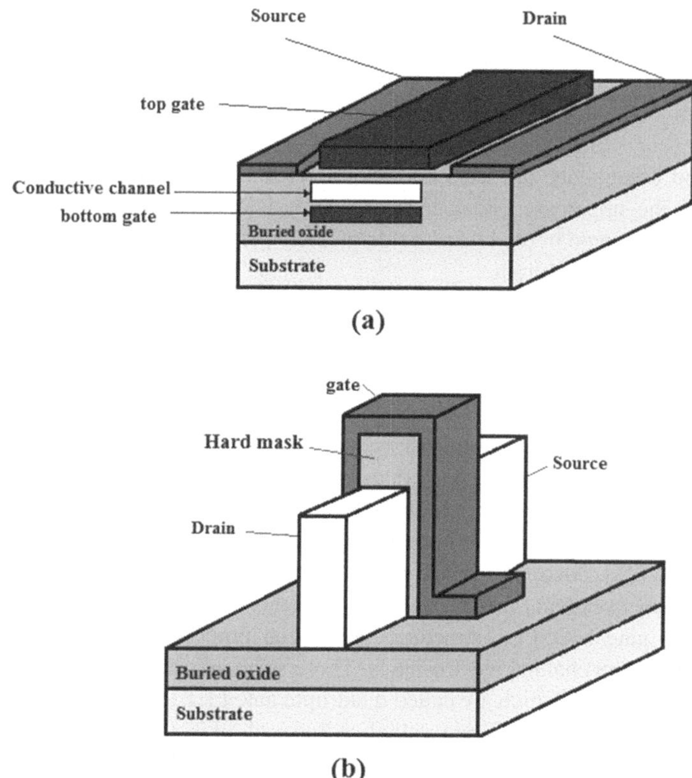

Fig. 2.13 Various flavors of double-gate (**a**) planar, (**b**) FinFET

of one gate can be modulated by the bias applied to another gate which uses for signal modulation.

Triple-Gate MOS Devices

The triple-gate transistor is a non-planar narrow MOS device wherein the silicon body is covered by the gate from three of its sides. Since the better electrostatic control of the channel is gained by using more gates, so that the silicon body thickness and width can be larger than for the FD-SOI and double-gate devices, respectively.

As shown in Fig. 2.14, the triple-gate structure is constituted of a thin semiconductor body stands on a substrate with a gate dielectric which is placed on top and sidewalls of the device body. Source and drain terminals are positioned on two ends of the semiconductor body. Since the semiconductor body is surrounded by gates on three sides, three separated channels are available for electrons to travel. The gate width of the transistor is equivalent to the sum of each three gates. To avoid several threshold voltages, the thickness of three gates' dielectric must be equal. For enhancing the current drive of the triple-gate device, one solution is to extend the side gates into the buried oxide on both borders. This favor of multi-gate devices is called pi-gate [26]. Another solution is to extend the side gates under the silicon body from both sides. These devices are called omega-gate [27]. The tri-gate transistor (Fig. 2.15a) is so named as the gate has three sides. The names of pi-gate and omega-gate are used because the cross-section views of these devices are like to the uppercase Greek letters of Π and Ω, respectively (Fig. 2.15b, c). Extension of the gate in the buried oxide protects the back of the channel area from the electric field of the drain.

Surrounding Gate Devices

Among the various multi-gate transistors, surrounding gate or gate-all-around (GAA) transistors attain the highest packing density due to their better current drive. In these devices, around the channel is enfolded entirely by gate oxide and the gate electrode. This structure provides the most improvement in transconductance and short-channel performance [28]. Also, surrounding gate MOSFETs have a high packing density owing to their excellent current drive over to other multi-gate MOSFETs. Several surrounding gate MOSFET structures have been proposed to more improve the engineering of the channel electrostatics. These structures are categorized in square cross-section devices which are called quadruple gate FET (Fig. 2.14a), and circular cross-section devices which are called cylindrical FET (Fig. 2.14b). The GAA MOSFETs can also be in vertical structure (Fig. 2.14c) [29].

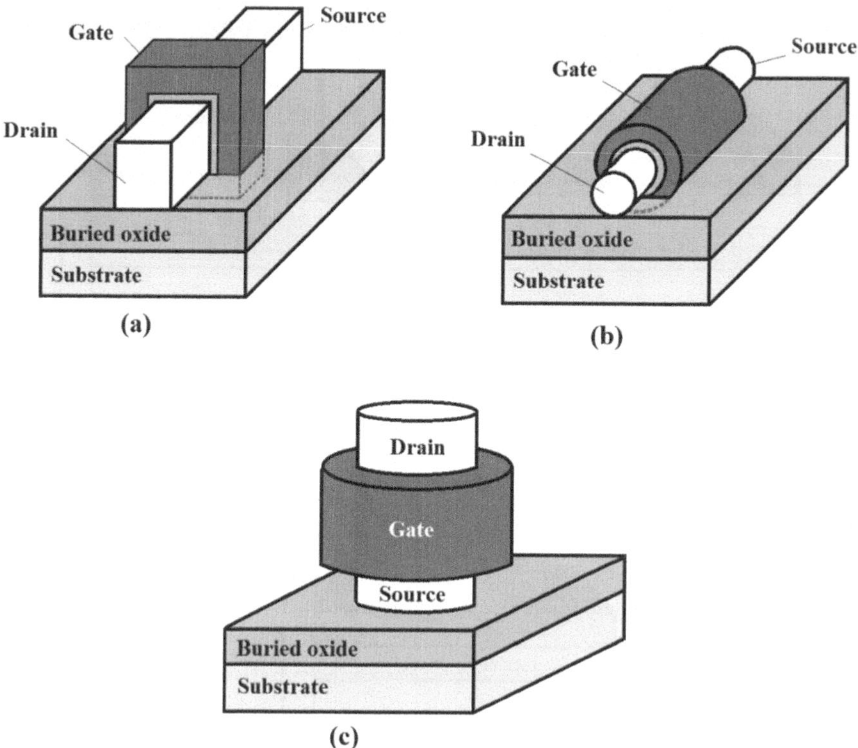

Fig. 2.14 Various kinds of surrounding gate devices: (**a**) quadruple-gate; (**b**) cylindrical FET; and (**c**) vertical GAA

Nanowire MOSFETs

If the cross body dimensions of a gate-all-around MOSFET are few nanometers, the device is called nanowire MOSFET. Since the cross-section area of the silicon body is only some nanometers, the device operates as a quantum wire, in which the carrier transportation between source and drain is one-dimensional. Therefore, the behavior of nanowire MOSFET is different from GAA which the movement of carriers in the body is in two spatial dimensions [30]. In this nanoscale device, according to the applied gate voltage, current flows through the nanowire or is pinched off. However, due to the small size of the device, single nanowires cannot carry sufficient current to make an effective transistor. Therefore, they are made as a transistor consists of multi nanowires that are under the control of a single gate and so operate as a single transistor (Fig. 2.16).

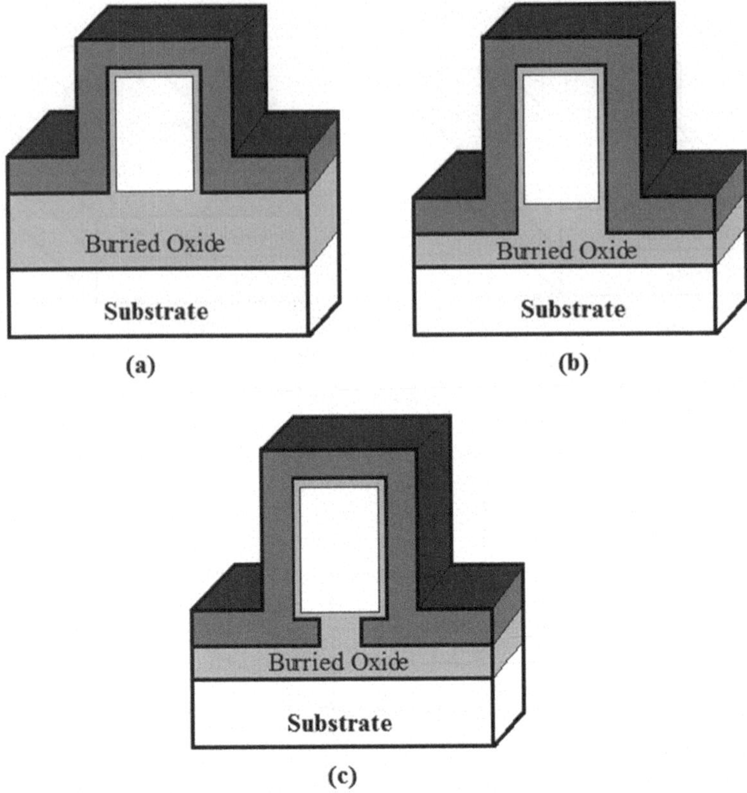

Fig. 2.15 The existing gate configurations for triple-gate MOS devices: (**a**) tri-gate; (**b**) pi-gate; and (**c**) omega-gate

2.7 Material-Based Non-CLASSICAL Devices

The use of other materials with better transport properties into MOS devices has the potential to increase the mobility of the carriers. Typically, the semiconductor material is silicon, although other semiconductors such as germanium (Ge), silicon-germanium alloys (SiGe), indium arsenide (InAs), and graphene can also be used. New materials can be used in the gate stack (dielectrics and electrode materials), conducting channel (to improve carrier transport), and also in the source and drain areas (to reduce resistance and improve carrier injection). In the 1990s, silicon, silicon dioxide, and aluminum were the primary materials used in typical MOS devices. Today, MOSFETs contain many other materials. Silicon-germanium has been used in the source–drain regions to strain the channel, with increasing ON-state drive current. Using dielectrics such as silicon oxynitride (SiON) and Hafnium oxide (HfO_2) with a higher permittivity than silicon dioxide (SiO_2) allows increasing the gate to channel capacitance without reducing the dielectric thickness. Also, various

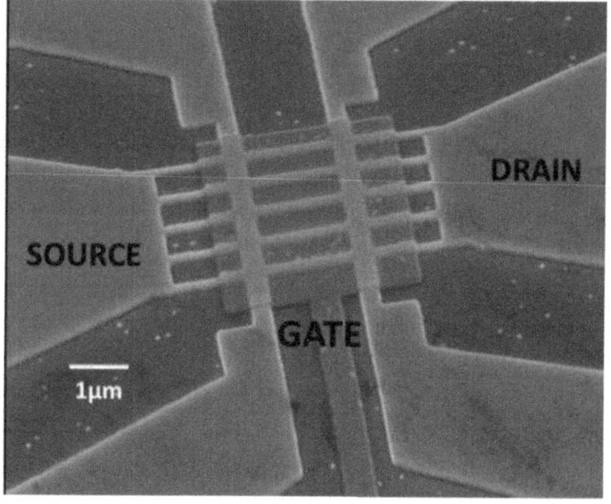

Fig. 2.16 Scanning electron microscope micrograph of multi nanowire field effect transistor with an array of five nanowires of width 40 nm

metal gates have been employed in lessening the gate depletion problem and in decreasing the resistant of the gate electrode. Also, introducing a high-k gate dielectric, low-k interconnects, and replacing of bulk silicon with strained silicon on insulator (sSOI) are another approaches that can be used in novel MOS devices. Partial silicidation of the source/drain contacts by cobalt, platinum, nickel, etc. has reduced series resistance of MOSFET, while copper and tungsten have replaced aluminum in the interconnect layers and via plugs [31].

2.7.1 *High-k Metal Gate Stack MOSFETs*

The dielectric constant (k) is a quantity of a material's potency to resist the formation of an electrical field inside it. When a voltage is applied to a material with low dielectric constant, almost no change in orientation of its molecules is seen. But materials with high dielectric constants polarize their structures as a dipole to counteract the field. Since the strength of the dipoles rises, a stronger arrangement of the dipoles will be obtained, which leads to an improved value of k (Fig. 2.17).

These aligned dipoles cause an image charge effect at the interface of the dielectric-doped silicon layer. If the thickness of gate oxide is scaled down below the tunneling limit, the leakage current of gate immensely increases. Physically thicker dielectric layers are required to avoid tunneling current. For thicker gate dielectric, a material with high dielectric constant is needed to preserve its electrical characteristics of the device. A more popular high-k dielectric material used in MOS devices is HfO_2 which has a dielectric constant about five times higher than SiO_2. The various high-k dielectrics and their dielectric constant are listed in Table 2.2.

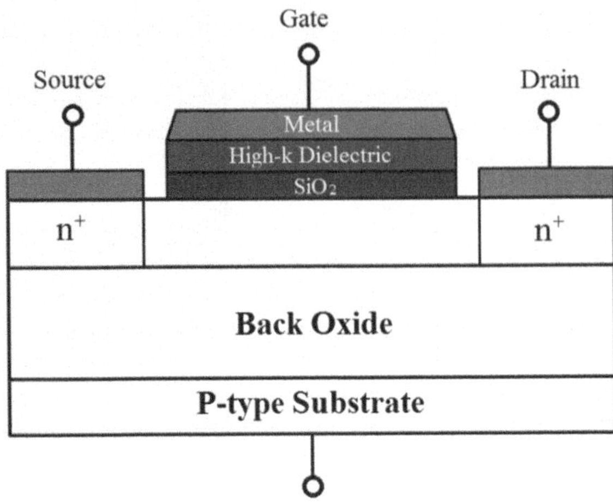

Fig. 2.17 The cross-section view of high-k metal gate stack MOSFETs

Table 2.2 High-k gate dielectric materials with their dielectric constants

Material	Symbol	Dielectric constant (*k*)
Silicon dioxide	SiO_2	3.9
Silicon nitride	Si_3N_4	7
Aluminum oxide	Al_2O_3	9
Yttrium(III) oxide	Y_2O_3	15
Lanthanum oxide	La_2O_3	30
Titanium dioxide	TiO_2	80
Hafnium(IV) oxide	HfO_2	25
Zirconium dioxide	ZrO_2	25
Tantalum pentoxide	Ta_2O_5	26

2.7.2 Strained MOSFETs

Strained MOSFET or SiGe MOSFET is a robust strategy to develop novel devices with enhanced functionalities. This device takes advantage of strain-induced enhancement of carrier transport in Si [32].

As shown in Fig. 2.18, strained MOSFET contains a strained Si channel grown on a layer of relaxed SiGe buffer. Due to the mismatch between two lattice constants, the growth of a crystal layer over another layer causes strain in the upper layer. Epitaxial relaxed SiGe layer forms a larger lattice constant on a silicon substrate, producing strain in tension or compression in SiGe layer. In this situation, the silicon is strained, because it has a higher lattice constant than the relaxed $Si_{1-x}Ge_x$ layer, where x is the fraction of germanium atoms [33]. From the viewpoint of theory, under either tensile or compressive strain conditions, the mobility of holes in silicon rises because the valence band breaks and conductivity mass reduces. Also, the mobility of electron is enhanced in silicon under the tensile strain because of the population of electron increases in two lower energy valleys with reduction of conductive mass.

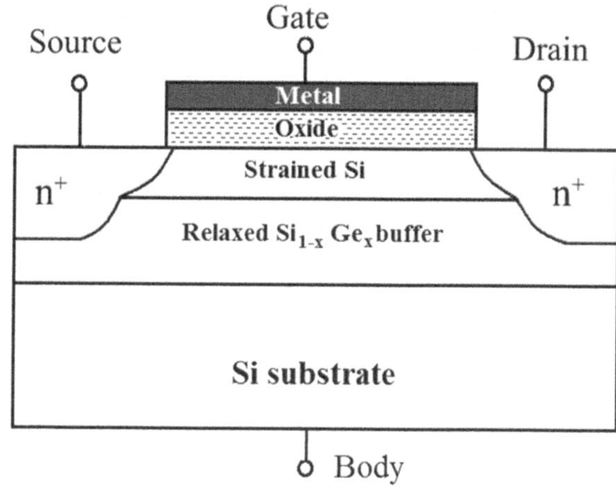

Fig. 2.18 Schematic representation of strained MOSFET

2.7.3 *Dual Material Gate (DMG) MOSFETs*

In these devices, as shown in Fig. 2.19, two different materials with different work-functions are laterally merged and form a single gate of the device. Using this technique enhances the gate transport efficiency by modifying the distribution of electric field and also surface potential profile along the conductive channel [34].

For n-channel DMG device, the work-function of metal gate 1 (M1) must be greater than metal gate 2 (M2). The difference between two work-functions creates a difference in the conforming flat-band voltages which brings a step potential profile. The step potential provides screening of the channel area under the metal gate adjacent to the source in the variations of drain potential. Consequently, the drain bias cannot affect the minimum potential of the channel. It has been demonstrated that DMG structure successfully suppresses SCE and DIBL [35].

2.8 Future Generation Devices

At the beginning of its fifth decade, the semiconductor technology continues to grow at a beautiful step. High-speed and low-power integrated circuits (IC) are used in an ever-increasing plenty of applications, permeating every part of human life. The string of significant advances that followed has led to today's gigahertz microprocessors and gigabit memories. The efficient downscaling of semiconductor devices and improvements in processing technologies have enabled us to attain extremely complex electronic systems. The scaling trend of MOS devices has been continued for more than four decades to achieve new semiconductor devices with higher packaging density, higher speed, and lower power dissipation. Accordingly, several non-classical structures with different approaches have been germinated to diminish SCEs, but each has its disadvantages as well. Consequently, the introducing of new

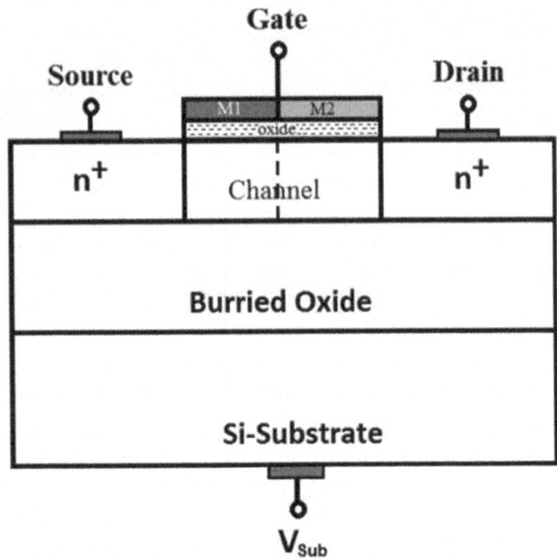

Fig. 2.19 Schematic cross-section view of dual material gate MOSFET

device structures which can preserve the ultimately scaled territory must be continued to minimize the problems related to these non-classical devices. The international technology roadmap for semiconductor (ITRS) proposed next-generation transistors includes the 3-D structure silicon devices, the new channel material devices, and also the gate stack material devices as displayed in Fig. 2.20. As described earlier, to reduce the difficulties associated with downscaling, SOI devices replaced the conventional bulk MOS devices. Introduction of more gates (multi-gate MOSFETs) offered even improved immunity to SCEs and also DIBL effects. Also, a different method, called gate material engineering was planned to provide more improvement of devices.

2.9 A Brief Review of the Previous Works

Due to the successive necessities for increasing current drive simultaneous with better suppression of SCEs during recent years, numerous studies have been down on non-classical silicon devices due to their better electrical properties and also their compatibility with typical silicon processing. This section is discussed in two parts. The first part provides a general overview of the previous works on the introducing, analysis, and modeling of MOSFETs during their progress from classical single-material single-gate to multi-material multi-gate. The second part belongs to the related works on silicon MESFETs.

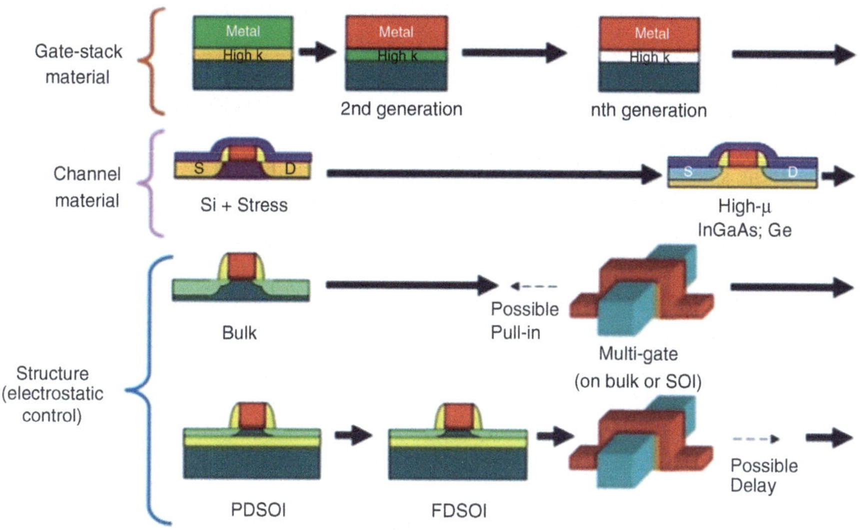

Fig. 2.20 Future semiconductor devices in the ITRS roadmap

2.9.1 *The Related Works on MOSFETs*

The concept of SOI dates back to 1964 where partially depleted (PD) devices fabricated on silicon-on-sapphire substrates [36]. In 1979, Toyabe and Asay [37] derived analytical models of the threshold voltage and breakdown voltage of short-channel MOSFETs based on the combination of approximate analytical solution and two-dimensional numerical analysis. It is shown that the potential function along the distance from the drain is an exponential function, and the threshold voltage exponentially decreases with decreasing of channel length. In 1980, Sano et al. [38] carried out an exact numerical analysis on MOSFETs fabricated on an epitaxial silicon layer that was grown on a buried oxide layers. In the same year, Worley [39] developed an enhanced model of the threshold voltage for the fully depleted SOS transistor. In 1982, Omura [40] proposed a channel potential profile model, in which the barrier lowering effect is taken into account, for a buried-channel SOI-MOSFET. The simple description for threshold voltage and drain breakdown voltage is found from the models. The charge coupling between the front and back gates of thin-film SOI-MOSFETs is evaluated, and closed-form model for the threshold voltage under all promising conditions is derived by Kiu-Lim et al. in 1983 [41]. After 1 year, they developed a simple mathematical model for the steady-state I–V characteristics of strongly inverted SOI-MOSFETs [42].

The first fully depleted SOI-MOSFET dates back to the middle 1980s where it was established that these devices showed excellent transconductance, current drive, and subthreshold swing [43]. In 1987, the double-gate SOI transistor as a new device with significantly enhanced performance, which forces the entire of the silicon film in strong inversion, is studied by Balestra et al. [44]. Two years later Young [45] examined the short-channel effect in fully depleted SOI-MOSFETs presenting a two-dimensional analytical model and also by computer simulation. The given model is used to calculate the potential function throughout the SOI layer and afterward to determine the threshold voltage reduction owing to the short-channel effect. In the same year, short-channel effects in thin-film SOI-MOSFETs are examined, mathematically and experimentally. Also, it is demonstrated that the short-channel effects depend on film thickness and body and also substrate biases [46].

In the year 1991, Chen and Huang [47] have offered an analytical model for threshold and subthreshold operation of SOI-MOSFET, considering the voltage drop across the substrate of the device. The analysis shows that both front-gate threshold voltage and subthreshold swing are degraded if the substrate effect is considered. In 1992, Yan et al. [48] proposed some guidelines for designing MOSFETs. They introduced a theoretical framework for both bulk and SOI structures which yields a natural length scale that can be successfully used as a design guideline. One year later, Suzuki et al. [49] established another scaling theory for double-gate SOI-MOSFETs which clarify how to define film thickness and oxide thickness for a particular gate length keeping a proper S-factor.

In 1994, Francis [50] presented an analytical model considering both the doping impurity charges and the electron concentration for ultra-thin double-gate n-MOS. A different assumption of the constant difference between surface and mid-film potentials is introduced. In the same year, Tosaka and Suzuki [51] derived a simple expression for the subthreshold swing in a double-gate SOI-MOSFET which relies only on a scaling device parameter.

In 1995, Suzuki et al. [52] suggested $n+-p+$ double-gate SOI-MOSFETs, and established the model for this device. In this year, a threshold voltage model for deep-submicron FD-SOI-MOSFET using a quasi-model is proposed by Banna et al. Moreover, they evaluated the statistical variation of threshold voltage versus silicon film thickness variations [53].

For the first time, Long et al. [54] reported a new bulk configuration, the dual material gate MOSFET, wherein two dissimilar materials with different work-functions, which laterally merged as a single gate, were used for the gate. This structure gave improving both the carrier transport efficiency and superior short-channel effects due to inducing a step-function in the surface potential.

Park et al. [26] described TCAD simulations of different SOI-MOSFETs with double and triple-gate structures, as well as gate-all-around devices. The concept of a triple-gate MOSFET with sidewalls extending into the buried oxide layer (as a result of this named a "Π-gate" or "pi-gate" MOSFET) is introduced and analyzed. In the year 2003, Suzuki and Pidin [55] derived an analytical model for fully depleted single-gate SOI-MOSFETs which has two-dimensional effects in both SOI and BOX layers. This model is valid for both long and short-channel SOI-MOSFETs and exhibits the dependence of short-channel effects on the device parameters.

The similar study has been down on the SOI-MOSFETs by Kumar and Chaudhry [35]. They demonstrated that the surface potential in the channel region is affected by step-function which provides the screening of the drain potential variation by the gate near the drain resulting in suppressed SCEs similar to the hot carrier effect (HCE) and drain-induced barrier lowering (DIBL). In 2004, Chaudhry and Kumar [56] studied the performance degradation of SOI devices due to the short-channel effects since the channel length is scaled down. A new device called the dual-material gate SOI-MOSFET is discussed, and its efficiency in suppressing SCEs is investigated. They also theoretically explored the features of a dual-material gate FD-SOI-MOSFET and compared the results with those of a similar SOI-MOSFET [57]. The two-dimensional TCAD simulation showed the enhancing of threshold voltage roll-up and is suppressing the short-channel effects. Furthermore, these devices can be controlled by engineering the work-function and length of the gate material.

In the same year, Ma and Kaya [58] studied the structure of double-gate SOI-MOSFETs. They also analyzed the impact of various physical mechanisms on device linearity. Colinge [59] described the evolution and the properties of various multiple-gate devices from single-gate to multi-gate transistors. Also, he introduced and analyzed a new class of MOSFETs named triple-plus-gate device. Such structures are triple-gate devices with field induced, the pseudo-fourth gate that imitates a real back gate, fourth gate, under the device. In 2005, the idea of the dual-material gate had been used for the DG SOI-MOSFET structure and new features, the dual material double gate (DMDG), have been introduced and examined for the first time by Reddy and Kumar [60]. They have validated that the DMDG structure leads to reduced SCEs and HCEs.

An analytical model for the subthreshold surface potential of deep submicron DMG MOSFETs is proposed by Baishya et al. [61]. This model takes into account the non-uniformity in the channel depletion layer depth owing to the source and drain junctions, and also to the work-function difference of the gate materials. In the year 2007, Hamid et al. [62] developed analytical physically based models for the threshold voltage, containing the DIBL effect, and the subthreshold swing of un-doped symmetrical double-gate MOSFETs.

One the same year, Yang et al. [63] established the scaling theory of Fin field effect transistors (FinFETs) by a 3-D analytical solution of Poisson's equation in the channel region and numerical simulation.

In 2008, Agarwal et al. [64] reported a closed-form solution of surface potential for all the three surfaces (gate oxide—silicon film interface, silicon film—buried oxide interface, and buried oxide—substrate interface) of FD-SOI-MOSFETs. In this model, it is supposed that the silicon film is always fully depleted and the back silicon film surface is never inverted. Also, the effect of substrate charge explicitly is considered.

In 2009, Chiang [65] offered a new analytical compact model for a short-channel tri-material gate stack SOI-MOSFET comprising 2-D channel potential, threshold voltage, and subthreshold current. Also, Jooyoung et al. [66] presented a more detailed description of compact modeling for various forms of multiple-gate MOSFETs. Ritzenthaler et al. [67] developed a model to gain analytical expressions of the subthreshold characteristics of pi-gate FET transistors. Based on the solution of the 3-D form of Laplace's equation, they described the interface coupling in the structure and calculated the surface potential and subthreshold current.

Meel [68] developed a 3-D analytical model of front and back gate threshold voltages for FD-SOI-MOSFETs. The present model includes all the three paths of conduction like front gate oxide/silicon, back gate oxide/silicon, and sidewall oxide/ silicon interfaces in a mesa-isolated structure of this device. Then the 3-D Poisson's equation has been solved analytically using proper boundary conditions to obtain an expression of potential function within silicon film. Also, Tsormpatzoglou et al. [69] analyzed and modeled the lightly doped tri-gate MOSFETs, using the superposition of the threshold voltages of a symmetric and an asymmetric double-gate MOSFET. They consider the TG MOSFET as a sum of two independent double-gate MOSFETs, a symmetric DG MOSFET which is located between the two lateral gates and an asymmetric one which located between the top and bottom gates. Accordingly, the analytical model for TG device is driven as a perimeter-weighted sum of analytical expressions for two DG MOSFETs. Kloes et al. [70] introduced a new compact drain-current model for double-gate or triple-gate SOI-MOSFETs, which is based on a physics-based 3-D analysis. They derived analytical model equations for the height of the potential barrier in closed form from a 3-D model for the channel electrostatics without the necessity to introduce any fitting parameter.

Vimala and Balamurugan [71] developed another analytical model for tri-gate MOSFETs considering quantum effects. The model is based on the mathematical solution of Schrodinger Poisson's equation using variation method. Tripathi and Narendar [72] presented a surface potential and subthreshold current model for doped FinFET following the separation of variable technique. The model displays the variation of channel potential versus the thickness of oxide layer and doping concentration.

2.9.2 The Related Works on Silicon MESFETs

Silicon MESFETs fabricated on bulk and SOI substrates have down notable attention for more than a decade, and their potency as a good contender in VLSI technology has been studied and investigated.

Marshall [73] found an analytical model that predicts small-geometry effects in Si MESFETs. The model is founded on a two-dimensional analytical solution of Poisson's equation in the subthreshold domain that applies to the junction-isolated structure classic of silicon devices. The authors decomposed the potential function as the sum of the two-dimensional solution to the Poisson's equation and one-dimensional solution to the Laplace's equation.

Hou and Wu [74] used the Green function method to develop an analytical model for the 2-D potential function and subthreshold behavior of SOI-MESFETs. Although the model was accurate, it includes a lot of complicated mathematical equations. Also, the method of defining the threshold voltage was not clear. Cao [75] simplified the model presented by Hou. The superiority of this model is that the

Fourier coefficient of the electric movement at the interface of Si–SiO2 is given in the closed form, rather than infinite series as in the prior model. However, the model is yet complicated and confusing for understanding.

Chiang et al. [76] found another analytical model for the same device using the superposition technique. The mathematical complexity of this model is impressively reduced in comparison to the last model, but it is reported that the resultant data from the model are different from that obtained from the graphs presented in the paper [77]. Pandey et al. [77] presented another model which is based on the assumption that the potential function is approximated as a parabolic potential function. Although this model is simple, it is not as accurate as the last model. Jit et al. [78] proposed modified boundary condition at the Si–SiO2 interface to remove the drawbacks related to the model presented by Chiang. They also studied and modeled the subthreshold current, and the subthreshold swing of FD-SOI-MESFETs considering the short-channel effects in another work [79]. The result shows that the subthreshold drain current of an SOI-MESFET is considerably reduced in comparison to GaAs MESFET. Moreover, subthreshold swing is minimized by increasing the back oxide thickness and also by reducing the silicon film thickness of the device. The last outstanding reported analytical model of SOI-MESFET is offered by Suh [80]. In this paper, the doping density of silicon film is considered as an arbitrary function of the vertical direction. Taking into account the bottom channel potential in the lateral direction and using the natural length expression, the 2-D potential function in both silicon film and buried oxide layer was obtained. This model can describe the threshold voltage of a short-channel SOI-MESFET regarding device parameters and applied voltages properly.

Hashemi et al. investigated the impacts of the dual metal gate on the FD-SOI-MESFETs by presenting a new 2-D analytical model for the bottom potential and threshold voltage for various device parameters [81]. The results of their study confirmed that using DMG in FD-SOI-MESFETs induces a step-function in the bottom potential profile in the channel too. This step-function decreases the drain-induced barrier lowering (DIBL) effect. Likewise, the effects of different DMG structure parameters on DIBL are shown. Chiang [82] examined the subthreshold behavior of SOI-MESFETs by presenting an analytical model. He found that the thin silicon film can alleviate short-channel effects and has a propensity to cause small threshold voltage roll-off and subthreshold swing degradation. Lakhdar et al. [83] present a new structure of GaN-MESFET named double-gate GaN-MESFET which offers enhancement in the screening of the channel potential variation, and an enhancement of the short-channel effects. They also examined the device performance by deriving an analytical model for channel potential, threshold voltage, and drain-induced barrier lowering (DIBL). However, no comparable study has been down to examine the multi-gate structure in the case silicon SOI-MESFET. Orouji et al. [84] introduced and analyzed a new SOI-MESFET structure by modifying the form of the buried oxide. In the proposed structure, a double protruded area with a groove in the buried oxide is shaped. This approach reduces the concentration of carriers in the channel and increases the breakdown voltage. Based on this comprehensive and detailed literature review, most of the work done so far has been on

MOSFETs. Still, SOI-MESFETs are beginning to found their place in the VLSI territory as a replacement to MOSFETs in silicon ICs. The silicon MESFET is a significant class of transistors used in electronic equipment and systems. Due to the applications of SOI-MESFETs in high-speed and high-power digital circuits, we approached for introducing and analyzing of non-classical silicon on insulator MESFET along with the 2-D and 3-D mathematical modeling and TCAD simulation.

References

1. K. Gupta, N. Gupta, *Advanced Semiconducting Materials and Devices* (Springer, Berlin, 2016)
2. J.B. Kuo, K.-W. Su, *CMOS VLSI Engineering: Silicon-on-Insulator (SOI)* (Springer, Berlin, 2013)
3. I. Bahl, *Fundamentals of RF and Microwave Transistor Amplifiers* (Wiley, Hoboken, 2009)
4. L.I. Berger, *Semiconductor Materials* (CRC Press, Boca Raton, 1996)
5. M. Golio, *RF and Microwave Semiconductor Device Handbook* (CRC Press, Boca Raton, 2017)
6. S. Henzler, *Power Management of Digital Circuits in Deep Sub-Micron CMOS Technologies* (Springer, Berlin, 2006)
7. S. Saurabh, M.J. Kumar, *Fundamentals of Tunnel Field-Effect Transistors* (CRC Press, Boca Raton, 2016)
8. S.M. Kang, Y. Leblebici, *CMOS Digital Integrated Circuits: Analysis and Design* (TMH, New Delhi, 2003)
9. F. D'Agostino, D. Quercia, Short-channel effects in MOSFETs, 2000. Accessed online http://www0.cs.ucl.ac.uk/staff/d.quercia/projects/vlsi/report.pdf
10. A.K. Singh, *Electronic Devices and Integrated Circuits* (PHI Learning Pvt. Ltd., New Delhi, 2011)
11. L. Wang, *Quantum Mechanical Effects on MOSFET Scaling Limit* (Citeseer, 2006)
12. L. Wilson, *International Technology Roadmap for Semiconductors (ITRS)* (Semiconductor Industry Association, Washington, 2013)
13. T. Skotnicki, J.A. Hutchby, T.-J. King, H.-S. Wong, F. Boeuf, The end of CMOS scaling: Toward the introduction of new materials and structural changes to improve MOSFET performance. IEEE Circuits Devices Mag. **21**, 16–26 (2005)
14. L. Hyunjin, L. Jongho, S. Hyungcheol, DC and AC characteristics of sub-50-nm MOSFETs with source/drain-to-gate nonoverlapped structure. IEEE Trans. Nanotechnol. **1**, 219–225 (2002)
15. G.G. Shahidi, in *2001 International Symposium on VLSI Technology, Systems, and Applications, Proceedings of Technical Papers*. SOI Technology for the GHz Era (IEEE, 2001), pp. 11–14
16. P. Feng, Design, Modeling and Analysis of Non-classical Field Effect Transistors, 2012
17. A. Kranti, S. Haldar, R. Gupta, Analytical model for threshold voltage and I–V characteristics of fully depleted short channel cylindrical/surrounding gate MOSFET. Microelectron. Eng. **56**, 241–259 (2001)
18. W. Ma, *Linearity Analysis of Single and Double-Gate Silicon-On-Insulator Metal-Oxide-Semiconductor-Field-Effect-Transistor* (Ohio University, Athens, 2004)
19. S. Shee, *Quantum Analytical Modeling of Ultrathin DMDG SON MOSFET: A Performance Assessment* (Jadavpur University, Kolkata, 2014)
20. J.B. Kuo, S.-C. Lin, *Low-Voltage SOI CMOS VLSI Devices and Circuits* (Wiley, Hoboken, 2004)

21. S. Crisoloveanu, S. Li, *Electrical Characterization of SOI Materials and Devices* (Kluwer Academic Publishers, Dordrecht, 1995)
22. P. Jong-Tae, J.P. Colinge, Multiple-gate SOI MOSFETs: device design guidelines. IEEE Trans. Electron Devices **49**, 2222–2229 (2002)
23. T. Sekigawa, Y. Hayashi, Calculated threshold-voltage characteristics of an XMOS transistor having an additional bottom gate. Solid State Electron. **27**, 827–828 (1984)
24. L. Hyung-Kyu, J.G. Fossum, Threshold voltage of thin-film silicon-on-insulator (SOI) MOSFET's. IEEE Trans. Electron Devices **30**, 1244–1251 (1983)
25. C. Mallikarjun, K.N. Bhat, Numerical and charge sheet models for thin-film SOI MOSFETs. IEEE Trans. Electron Devices **37**, 2039–2051 (1990)
26. J.-T. Park, J.-P. Colinge, C.H. Diaz, Pi-gate SOI MOSFET. IEEE Electron Device Lett. **22**, 405–406 (2001)
27. H.-W. Gao, T.K. Chiang, in *2013 IEEE 10th International Conference on ASIC (ASICON)*. A Novel Scaling Theory For Fully-depleted Omega-gate (ΩG) MOSFETs (IEEE, 2013), pp. 1–3
28. S. Jae Young, C. Woo Young, P.J. Hee, L. Jong Duk, P. Byung-Gook, Design optimization of gate-all-around (GAA) MOSFETs. IEEE Trans. Nanotechnol. **5**, 186–191 (2006)
29. B. Yang, K.D. Buddharaju, S.H.G. Teo, N. Singh, G.Q. Lo, D.L. Kwong, Vertical silicon-nanowire formation and gate-all-around MOSFET. IEEE Electron Device Lett. **29**, 791–794 (2008)
30. F. Schwierz, J.J. Liou, H. Wong, *Nanometer CMOS* (Pan Stanford, 2010)
31. G. Hellings, K. De Meyer, *High Mobility and Quantum Well Transistors* (Springer, Berlin, 2013)
32. R. Ismail, M.T. Ahmadi, S. Anwar, *Advanced Nanoelectronics* (CRC Press, Boca Raton, 2012)
33. A. Chaudhry, J. Roy, G. Joshi, Nanoscale strained-Si MOSFET physics and modeling approaches: A review. J. Semicond. **31**, 104001 (2010)
34. D. Yuehua, H. Yuan, Q. Liu, K. Daoming, C. Junning, in *IEEE Asia Pacific Conference on Circuits and Systems, 2006. APCCAS 2006*. Physics-based Modeling and Simulation of Dual Material Gate (DMG) LDMOS (IEEE, 2006), pp. 1500–1503
35. M.J. Kumar, A. Chaudhry, Two-dimensional analytical modeling of fully depleted DMG SOI MOSFET and evidence for diminished SCEs. IEEE Trans. Electron Devices **51**, 569–574 (2004)
36. C.W. Mueller, P.H. Robinson, Grown-film silicon transistors on sapphire. Proc. IEEE **52**, 1487–1490 (1964)
37. T. Toyabe, S. Asai, Analytical models of threshold voltage and breakdown voltage of short-channel MOSFET's derived from two-dimensional analysis. IEEE Trans. Electron Devices **26**, 453–461 (1979)
38. E. Sano, R. Kasai, K. Ohwada, H. Ariyoshi, A two-dimensional analysis for MOSFET's fabricated on buried SiO2 layer. IEEE Trans. Electron Devices **27**, 2043–2050 (1980)
39. E.R. Worley, Theory of the fully depleted SOS/MOS transistor. Solid State Electron. **23**, 1107–1111 (1980)
40. Y. Omura, A simple model for short-channel effects of a buried-channel MOSFET on the buried insulator. IEEE Trans. Electron Devices **29**, 1749–1755 (1982)
41. L. Hyung-Kyu, J.G. Fossum, Threshold voltage of thin-film silicon-on-insulator (SOI) MOSFET's. IEEE Trans. Electron Devices **30**, 1244–1251 (1983)
42. L. Hyung-Kyu, J.G. Fossum, Current-voltage characteristics of thin-film SOI MOSFET's in strong inversion. IEEE Trans. Electron Devices **31**, 401–408 (1984)
43. J.P. Colinge, Subthreshold slope of thin-film SOI MOSFET's. IEEE Electron Device Lett. **7**, 244–246 (1986)
44. F. Balestra, S. Cristoloveanu, M. Benachir, J. Brini, T. Elewa, Double-gate silicon-on-insulator transistor with volume inversion: A new device with greatly enhanced performance. IEEE Electron Device Lett. **8**, 410–412 (1987)
45. K.K. Young, Short-channel effect in fully depleted SOI MOSFETs. IEEE Trans. Electron Devices **36**, 399–402 (1989)

46. S. Veeraraghavan, J.G. Fossum, Short-channel effects in SOI MOSFETs. IEEE Trans. Electron Devices **36**, 522–528 (1989)
47. H.T. Chen, R.S. Huang, An analytical model for back-gate effects on ultrathin-film SOI MOSFETs. IEEE Electron Device Lett. **12**, 433–435 (1991)
48. R.H. Yan, A. Ourmazd, K.F. Lee, Scaling the Si MOSFET: From bulk to SOI to bulk. IEEE Trans. Electron Devices **39**, 1704–1710 (1992)
49. K. Suzuki, T. Tanaka, Y. Tosaka, H. Horie, Y. Arimoto, Scaling theory for double-gate SOI MOSFET's. IEEE Trans. Electron Devices **40**, 2326–2329 (1993)
50. P. Francis, A. Terao, D. Flandre, F. Van de Wiele, Modeling of ultrathin double-gate nMOS/SOI transistors. IEEE Trans. Electron Devices **41**, 715–720 (1994)
51. Y. Tosaka, K. Suzuki, T. Sugii, Scaling-parameter-dependent model for subthreshold swing S in double-gate SOI MOSFET's. IEEE Electron Device Lett. **15**, 466–468 (1994)
52. K. Suzuki, T. Sugii, Analytical models for n^+-p^+ double-gate SOI MOSFET's. IEEE Trans. Electron Devices **42**, 1940–1948 (1995)
53. S.R. Banna, M. Chan, P.K. Ko, C.T. Nguyen, C. Mansun, Threshold voltage model for deep-submicrometer fully depleted SOI MOSFET's. IEEE Trans. Electron Devices **42**, 1949–1955 (1995)
54. W. Long, H. Ou, J.M. Kuo, K.K. Chin, Dual-material gate (DMG) field effect transistor. IEEE Trans. Electron Devices **46**, 865–870 (1999)
55. K. Suzuki, S. Pidin, Short-channel single-gate SOI MOSFET model. IEEE Trans. Electron Devices **50**, 1297–1305 (2003)
56. A. Chaudhry, M.J. Kumar, Controlling short-channel effects in deep-submicron SOI MOSFETs for improved reliability: A review. IEEE Trans. Device Mater. Reliab. **4**, 99–109 (2004)
57. A. Chaudhry, M.J. Kumar, Investigation of the novel attributes of a fully depleted dual-material gate SOI MOSFET. IEEE Trans. Electron Devices **51**, 1463–1467 (2004)
58. W. Ma, S. Kaya, Impact of device physics on DG and SOI MOSFET linearity. Solid State Electron. **48**, 1741–1746 (2004)
59. J.-P. Colinge, Multiple-gate SOI MOSFETs. Solid State Electron. **48**, 897–905 (2004)
60. G.V. Reddy, M.J. Kumar, A new dual-material double-gate (DMDG) nanoscale SOI MOSFET-two-dimensional analytical modeling and simulation. IEEE Trans. Nanotechnol. **4**, 260–268 (2005)
61. S. Baishya, A. Mallik, C.K. Sarkar, A pseudo two-dimensional subthreshold surface potential model for dual-material gate MOSFETs. IEEE Trans. Electron Devices **54**, 2520–2525 (2007)
62. H.A.E. Hamid, J.R. Guitart, B. Iniguez, Two-dimensional analytical threshold voltage and subthreshold swing models of undoped symmetric double-gate MOSFETs. IEEE Trans. Electron Devices **54**, 1402–1408 (2007)
63. W. Yang, Y. Zhiping, T. Lilin, Scaling theory for FinFETs based on 3-D effects investigation. IEEE Trans. Electron Devices **54**, 1140–1147 (2007)
64. P. Agarwal, G. Saraswat, M.J. Kumar, Compact surface potential model for FD SOI MOSFET considering substrate depletion region. IEEE Trans. Electron Devices **55**, 789–795 (2008)
65. T.-K. Chiang, A new two-dimensional analytical subthreshold behavior model for short-channel tri-material gate-stack SOI MOSFET's. Microelectron. Reliab. **49**, 113–119 (2009)
66. S. Jooyoung, Y. Bo, Y. Yu, T. Yuan, A review on compact modeling of multiple-gate MOSFETs. IEEE Trans. Circuits Syst. I Regul. Pap. **56**, 1858–1869 (2009)
67. R. Ritzenthaler, F. Lime, O. Faynot, S. Cristoloveanu, B. Iñiguez, 3D analytical modelling of subthreshold characteristics in vertical multiple-gate FinFET transistors. Solid State Electron. **65–66**, 94–102 (2011)
68. K. Meel, R. Gopal, D. Bhatnagar, Three-dimensional analytic modelling of front and back gate threshold voltages for small geometry fully depleted SOI MOSFET's. Solid State Electron. **62**, 174–184 (2011)
69. A. Tsormpatzoglou, D.H. Tassis, C.A. Dimitriadis, G. Ghibaudo, N. Collaert, G. Pananakakis, Analytical threshold voltage model for lightly doped short-channel tri-gate MOSFETs. Solid State Electron. **57**, 31–34 (2011)

70. A. Kloes, M. Schwarz, T. Holtij, MOS^3: A new physics-based explicit compact model for lightly doped short-channel triple-gate SOI MOSFETs. IEEE Trans. Electron Devices **59**, 349–358 (2012)
71. P. Vimala, N.B. Balamurugan, New analytical model for nanoscale tri-gate SOI MOSFETs including quantum effects. IEEE J. Electron Devices Soc. **2**, 1–7 (2014)
72. S. Tripathi, V. Narendar, A three-dimensional (3D) analytical model for subthreshold characteristics of uniformly doped FinFET. Superlattice. Microst. **83**, 476–487 (2015)
73. J.D. Marshall, J.D. Meindl, An analytical two-dimensional model for silicon MESFETs. IEEE Trans. Electron Devices **35**, 373–383 (1988)
74. H. Chin-Shan, W. Ching-Yuan, A 2-D analytic model for the threshold-voltage of fully depleted short gate-length Si-SOI MESFETs. IEEE Trans. Electron Devices **42**, 2156–2162 (1995)
75. J.G. Cao, A simplified 2-D analytic model for the threshold-voltage of fully depleted short gate-length Si-SOI MESFET's. IEEE Trans. Electron Devices **43**, 1314–1315 (1996)
76. T.K. Chiang, Y.H. Wang, M.P. Houng, Modeling of threshold voltage and subthreshold swing of short-channel SOI MESFET's. Solid State Electron. **43**, 123–129 (1999)
77. P. Pandey, B.B. Pal, S. Jit, A new 2-D model for the potential distribution and threshold voltage of fully depleted short-channel Si-SOI MESFETs. IEEE Trans. Electron Devices **51**, 246–254 (2004)
78. S. Jit, P. Pandey, A. Kumar, S.K. Gupta, Modified boundary condition at Si–SiO2 interface for modeling of threshold voltage and subthreshold swing of short-channel SOI MESFET's. Solid State Electron. **49**, 141–143 (2005)
79. S. Jit, P.K. Pandey, P.K. Tiwari, Modeling of the subthreshold current and subthreshold swing of fully depleted short-channel Si–SOI-MESFETs. Solid State Electron. **53**, 57–62 (2009)
80. C.H. Suh, Analytical model for deriving the threshold voltage of a short gate SOI MESFET with vertically non-uniformly doped silicon film. Circuits Devices Syst. IET **4**, 525–530 (2010)
81. P. Hashemi, A. Behnam, E. Fathi, A. Afzali-Kusha, M. El Nokali, 2-D modeling of potential distribution and threshold voltage of short channel fully depleted dual material gate SOI MESFET. Solid State Electron. **49**, 1341–1346 (2005)
82. T.K. Chiang, in *9th International Conference on Solid-State and Integrated-Circuit Technology, 2008. ICSICT 2008*. The New Analytical Subthreshold Behavior Model For Dual Material Gate (DMG) SOI MESFET (2008), pp. 288–289
83. N. Lakhdar, F. Djeffal, A two-dimensional analytical model of subthreshold behavior to study the scaling capability of deep submicron double-gate GaN-MESFETs. J. Comput. Electron. **10**, 382–387 (2011)
84. A.A. Orouji, Z. Ramezani, S.M. Sheikholeslami, A novel SOI-MESFET structure with double protruded region for RF and high voltage applications. Mater. Sci. Semicond. Process. **30**, 545–553 (2015)

Chapter 3
Modeling of Classical SOI MESFET

3.1 Introduction

Since the advent of the first integrated circuit, the downscaling trend has led to extensive progress regarding cost, performance, and level of integration. It is well known that development in the level of integration is achieved by continuous reduction of the minimum size of electronic components. However, since downscaling has encroached on submicron territory, some undesirable effects have begun to pull down the performance of transistors. Also, the technological problems related to shrinking junctions and growing high-integrity thin oxide layer are apparent [1]. Hence, the reduction in physical dimensions needs to be concurrent with some other alterations like increasing of doping levels, reducing of insulator thickness, and decreasing of junction depth to degrade the second-order effects related to downscaling [2]. When gate lengths are scaled and doping levels are increased, electric fields tend to rise. The higher electric field, in small geometry devices, emitted the energetic carriers into the gate oxide layer and caused a shift in threshold voltage and therefore the reduction of transconductance with time (hot carrier effect). This problem can be lessened by reducing the junction electric field which is obtained by the reduction of the source and drain doping concentrations. But, low-doped regions, especially in small geometry devices, bring some other disadvantages like the high contact resistance. Also, lightly doped drain (LDD) MOS devices are introduced to repress the hot carrier effects. In these devices, introducing a narrow, lightly doped n-type region between the channel and the drain and source areas, the doping profile of the drain and source is modified. As a result, the electric field in the pinch-off region and thus the hot carrier effect and also the impact ionization are reduced. However, LDD MOSFETs have some drawbacks including the degradation of the current drive due to the increasing resistance and more processing complexity of the device manufacturing [3]. Another challenge of dimension downscaling relates to the ultra-thin oxide layer. The growing of high-quality and low-defect thin oxides is difficult. Moreover, quantum mechanical direct tunneling

I. S. Amiri et al., *Device Physics, Modeling, Technology, and Analysis for Silicon MESFET*, https://doi.org/10.1007/978-3-030-04513-5_3

(QMDT) is an important disadvantage in the ultra-thin gate oxide which remarkably reduces the device performance [4]. The last challenge is related to the junction depth scaling. The fact is that due to the various technological reasons, the junction depth has not been scaled pretty with scaling of channel lengths.

The leading technology used in modern VLSI circuits is based on the MOSFETs which are the most common transistors used in integrated circuits. However, due to the lack of oxide layer, MESFETs are free of the MOSFET-related problems. They naturally offer low-resistivity gate interconnect and bulk mobility [5]. MESFETs have no radiation sensitivity under the gate. They also offer higher channel mobility as the carriers can move further inside the silicon. Since the interaction between carriers and interface oxide is little, the noise of MESFETs is negligible [6]. Furthermore, MESFETs are the primary device for low-voltage applications, which is a favorite feature of most advanced VLSI technologies.

Due to the better scalability of Si MESFETs, these devices are being gradually considered as a replacement to MOSFETs for VLSI circuits [7]. These devices also offer enhanced radiation hardness, immunity to hot carrier aging, and less mobility degradation. Si MESFETs can be fabricated either on silicon bulk or on SOI substrate. The buried oxide diminishes the capacitive coupling to the substrate resulting in high-frequency parameters, low junction capacitance, latch-up immunity, improved device isolation [8], and also capability of operating at high temperatures [9]. These advantages make MESFETs appropriate for many electrical applications.

Unlike the MOSFETs that are tightly dependent on the device parameters in miniaturization, MESFETs are at a high degree of freedom. Silicon-based MESFETs are particularly promising technology for VLSI because they require potentially simpler process than GaAs MESFETs or silicon n-MOS [10].

The proper eliminating of the isolation problems in the SOI technology makes SOI MESFETs better than GaAs and Si-MESFETs for MESFET-based digital VLSI circuits and systems. The invention of complementary Si-MESFET (CMES) technology, as an alternative to the typical CMOS technology, offers better power consumption, circuit flexibility, and speed performance [11]. Finally, Si-MESFETs have the potential of being compatible with conventional CMOS technology. The use of advanced SOI CMOS procedure in the fabrication of SOI MESFET devices enriches the competency of the device for upcoming VLSI/ULSI systems.

3.2 Description of Device Structure

The cross-sectional view of a classical SOI MESFET used for analysis and modeling is shown in Fig. 3.1.

In the figure, x and y respectively indicate the longitudinal axis from the gate-edged source toward the drain and the transversal axis from the gate–silicon interface toward the buried oxide layer. L is the physical gate length, t_{si} is the thickness of

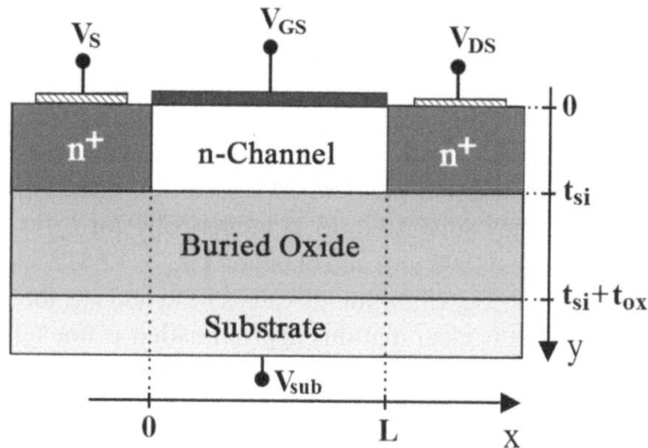

Fig. 3.1 Schematic cross-section view of classical SOI MESFET

silicon layer, and t_{ox} is the thickness of buried oxide layer. Also, V_{GS}, V_{DS}, and V_{Sub} represent the applied voltages to the gate, drain, and substrate terminals, respectively.

3.3 Definition of the Problem

For understanding and optimizing the electrical behavior of SOI MESFET devices with physical dimensions in the far submicron territory, it is necessary to develop a simple, reliable, and physically based analytical model which can describe the subthreshold performance of the device. Several analytical models have been developed to describe the electrical performance of this device. Most of these models are based on the solution of 2-D Poisson's equation in the depletion region under the gate. Many mathematical techniques including using Green function, the approximation of 2-D potential by a parabolic function [12, 13], and superposition method [14] are used to solve Poisson's equation. Green function method involves elaborate mathematical descriptions which are difficult to calculate. Approximation of potential distribution by a parabolic function leads to simple and less accurate model. Superposition method, which is more precise than other methods, has been used by Marshal et al. for silicon MESFET [15] and by Chiang et al. for SOI MESFET [16]. It is reported that the solution technique used by Chiang et al. has drawbacks that led to imprecise results [13]. They used the superposition method and found that the eigenvalue k_n satisfies the following equation:

$$\cot\left(k_n t_{si}\right) = -\frac{c_{ox}}{c_{si}} \frac{1}{k_n t_{si}} \quad (3.1)$$

They have used an approximation to Eq. (3.1) as:

$$k_n t_{\text{si}} \approx \frac{n\pi}{2} \qquad n = 1, 2, 3, \ldots, \infty \tag{3.2}$$

Using this approximation, they obtained the Fourier coefficients of the channel potential, but Jit et al. stated that this estimate is reasonable for $t_{\text{si}} << t_{\text{ox}}$ (i.e.$(c_{\text{ox}}/c_{\text{si}}) \to 0$) and is not accurate enough for all values of t_{si} and t_{ox} [2]. To prove this assertion, they solved Eq. (3.1) numerically for $t_{\text{ox}} = 0.2$ μm and two values of t_{si} ($t_{\text{si}} = 0.01$ and 0.08 μm) and obtained $k_1 t_{\text{si}} = 1.5812$ and $k_1 t_{\text{si}} = 1.6501$, respectively. According to Eq. (3.2), for all values of t_{si} and t_{ox}, the value of $k_1 t_{\text{si}}$ is approximated to 1.5708. It is clear that this approximation is not sufficiently precise. The discontinuity of Cotangent function at $k_n t_{\text{si}} = \frac{n\pi}{2}$ for even integer values of n is another drawback which leads to an undefined answer. Moreover, the property of orthogonality of eigenfunctions is ignored. Jit et al. offered solutions to resolve the mentioned problems. They ignored the effect of vertical electric field and modified the boundary condition at Si–SiO_2 interface as follows:

$$\left.\frac{\partial U(x,y)}{\partial y}\right|_{y=t_{si}} = 0 \tag{3.3}$$

Using this boundary condition, they obtained

$$k_n t_{\text{si}} \approx (2n-1)\frac{\pi}{2} \qquad n = 1, 2, 3, \ldots, \infty \tag{3.4}$$

Although in Eq. (3.4), the discontinuity of $\cot(k_n t_{\text{si}})$ for even values of n is eliminated, but this equation is not sufficiently precise. Computation of $k_1 t_{\text{si}}$ using Eq. (3.4) still results in an amount of 1.5708 for all values of t_{si} and t_{ox}. Therefore, this equation is not a general solution for all values of t_{si} and t_{ox}. Since the effect of the vertical electric field is ignored, the obtained model is not accurate enough for thin oxide SOI MESFETs. In this section, a new methodology is used to find k_n and the coefficients of Fourier series. An analytical model for channel potential, threshold voltage, and subthreshold swing of SOI MESFETs is derived. Making use of the model and simulation data, the subthreshold performance of the device under various device parameters and bias conditions is studied and discussed.

3.3.1 Calculation of Channel Potential

Just before the onset of high inversion, the potential profile in the silicon film is described by the Poisson's equation. In short-channel single-gate devices, the channel potential profile is purely two dimensional [17] which is defined as:

$$\frac{\partial^2 \Phi(x,y)}{\partial x^2} + \frac{\partial^2 \Phi(x,y)}{\partial y^2} = \frac{q}{\varepsilon_{si}}(N_A - N_D + n - p) \tag{3.5}$$

where $\Phi(x, y)$ is the potential at a point (x, y) in the silicon film, q is the charge of an electron, ε_{si} is the dielectric constant of silicon, and N_A and N_D are the uniform donor and acceptor doping density, respectively. Supposing that the impurity density of the channel is uniform, and n and p are the density of electron and hole. In the subthreshold region, the channel is supposed to be completely depleted of mobile carriers. So, we have $n, p << N_A$. Moreover, as the channel is n-type, we can write $N_A << N_D$. Accordingly, Eq. (3.5) is reduced to:

$$\frac{\partial^2 \Phi(x,y)}{\partial x^2} + \frac{\partial^2 \Phi(x,y)}{\partial y^2} = -\frac{qN_D}{\varepsilon_{si}} \qquad 0 \le x \le L \qquad 0 \ll y \ll t_{si} \tag{3.6}$$

The rectangle area in Fig. 3.1 shows the position of the channel region. To attain the solution of the 2-D Poisson's equation in the channel area, we can get discrete $\Phi(x, y)$ into two parts [16, 18] given by:

$$\Phi(x,y) = V(y) + U(x,y) \tag{3.7}$$

where $V(y)$ is the one-dimensional solution for long-channel circumstance, and $U(x, y)$ is the two-dimensional solution for the short-channel condition which accounts for the short-channel effects.

Boundary Conditions

To derive an analytical model for electrostatic potential function in the active channel (silicon layer), the Poisson's equation must be mathematically solved using the proper set of boundary conditions. Accordingly, we make use of four boundary conditions as follows.

Top Boundary

The top boundary is the border between the gate metal and the active channel. In the subthreshold region, the drain current is so small that the ohmic drop across under gate regions can be neglected. Accordingly, the surface potential under gate metal can be expressed as:

$$\Phi(x,y)\big|_{y=0} = V_{GS} - \Phi_{bi} \tag{3.8}$$

Here, V_{GS} is the applied gate-to-source voltage, and Φ_{bi} is the built-in potential of the Schottky barrier formed by the metal and the silicon film which is defined as

$\Phi_{\mathrm{bi}} = \Phi_{\mathrm{bn}} - V_n$. Hence, Φ_{Bn} indicates the barrier height of the gate metal–semiconductor junction and V_n is the potential difference between the conduction band and the Fermi level of the silicon film.

Bottom Boundary

The bottom boundary is the border between SOI layer and buried oxide layer. The electric flux at this interface is continuous:

$$\left.\frac{\partial \Phi(x,y)}{\partial y}\right|_{y=t_{\mathrm{si}}} = \frac{\varepsilon_{\mathrm{ox}}}{\varepsilon_{\mathrm{si}}} \frac{V_{\mathrm{sub}} - V_{\mathrm{fb}} - \Phi(x,y)|_{y=t_{\mathrm{si}}}}{t_{\mathrm{ox}}} \tag{3.9}$$

where $\varepsilon_{\mathrm{ox}}$ is the oxide permittivity, t_{si} is the channel thickness, t_{ox} is the buried oxide thickness, V_{sub} is the substrate voltage, and V_{fb} is the flat-band voltage of the substrate which accounts for the work-function difference between the substrate and silicon film. The flat-band voltage related to the back surface is given by:

$$V_{\mathrm{fb}} = \Phi_{\mathrm{sub}} - \Phi_{\mathrm{F}} \tag{3.10}$$

where $\Phi_{\mathrm{sub}} = \frac{KT}{q} . \ln\left(\frac{N_{\mathrm{sub}}}{n_i}\right)$ represents the substrate Fermi potential, k is Boltzmann's constant, T is the absolute temperature, N_{Sub} is the substrate doping concentration, and n_i is the silicon intrinsic carrier density.

Left Boundary

The left boundary is the left side of the depletion layer (source). The potential at the source–channel junction is taken as the built-in potential of an $n^{+}p^{+}$ junction and can be expressed as:

$$\Phi(x,y)|_{x=0} = V_{\mathrm{bi}} \tag{3.11}$$

V_{bi} is the built-in potential between the source (drain) and the undoped substrate. The built-in potential is calculated from the following formula:

$$V_{\mathrm{bi}} = \frac{E_g}{2} - \frac{kT}{q} \ln\left(\frac{N_D}{n_i}\right) \tag{3.12}$$

Hence, E_g is the energy gap of silicon.

Table 3.1 Boundary conditions for short-channel classical SOI MESFET

Boundary	Position	Potential	Electric field
Top	$y = 0$	$\Phi(x, 0) = V_{GS} - \Phi_{bi}$	–
Left	$x = 0$	$\Phi(0, y) = V_{bi}$	–
Right	$x = L$	$\Phi(L, y) = V_{bi}+V_{DS}$	–
Bottom	$y = t_{si}$	–	$\frac{\partial\Phi(x,t_{si})}{\partial y} = \frac{\varepsilon_{ox}}{\varepsilon_{si}} \frac{V_{sub} - V_{fb} - \Phi(x,t_{si})}{t_{ox}}$

Right Boundary

The right boundary is the right side of the depletion layer (drain). The proper boundary condition is similar to the left boundary condition, but the drain voltage is added to the built-in potential:

$$\Phi(x, y)|_{x=L} = V_{bi} + V_{DS} \tag{3.13}$$

where V_{DS} is drain-to-source voltage.

The boundary conditions for short-channel classical SOI MESFET are summarized in Table 3.1.

One-Dimensional Solution

Here, $V(y)$ is the solution of Poisson's equation in one dimension along the x-axis at a depth y in the SOI region of MESFET device:

$$\frac{d^2V(y)}{dy} = -\frac{qN_D}{\varepsilon_{si}} \qquad 0 \le y \le t_{si} \tag{3.14}$$

Regarding Eq. (3.7), the boundary conditions specified for $\Phi(x, y)$ must also be decomposed into two parts for the solution of $V(y)$ and $U(x, y)$. In doing so, we set the following boundary conditions for Eq. (3.14):

$$V(y)|_{y=0} = V_{GS} - \Phi_{bi} \tag{3.15}$$

$$\left.\frac{dV(y)}{\partial y}\right|_{y=t_{si}} = \frac{\varepsilon_{ox}}{\varepsilon_{si}} \frac{V_{sub} - V_{fb} - V(y)|_{y=t_{si}}}{t_{ox}} \tag{3.16}$$

These boundary conditions are taken from Eqs. (3.8) and (3.9). Now, integrating (3.14) twice concerning y, we obtain

$$V(y) = -\frac{qN_D}{2\varepsilon_{si}}y^2 + cy + d \tag{3.17}$$

Substituting boundary conditions (3.15) and (3.16) into (3.17), the arbitrary constants c and d are determined as:

$$c = \frac{(\gamma/t_{si})(\acute{V}_{sub} - \acute{V}_{GS}) + (qN_D/c_{si})(1+\gamma/2)}{1+\gamma} \tag{3.18}$$

$$d = V_{GS} - \Phi_{bi} \tag{3.19}$$

The quantities $c_{si} = \varepsilon_{si}/t_{si}$ and $c_{ox} = \varepsilon_{ox}/t_{ox}$ indicate the capacitances of the silicon film and buried oxide, respectively. Also, to make the expression more straightforward, the following constants are defined:

$$\acute{V}_{GS} = V_{GS} - \Phi_{bi} \tag{3.20}$$

$$\acute{V}_{sub} = V_{sub} - V_{fb} \tag{3.21}$$

$$\gamma = c_{ox}/c_{si} \tag{3.22}$$

Two-Dimensional Solution

Since the channel length is decreased to short, to investigate the dependence of device performance on its physical parameters, a Poisson's equation in two-dimensional form must be considered. On determining $V(y)$ from Eq. (3.14), $U(x, y)$ must satisfy the Laplace's equation as:

$$\frac{\partial^2 U(x,y)}{\partial x^2} + \frac{\partial^2 U(x,y)}{\partial y^2} = 0 \qquad 0 \le x \le L \qquad 0 \ll y \ll t_{si} \tag{3.23}$$

According to the specified boundary condition for $V(y)$, the resultant boundary conditions to Eq. (3.23) are defined as follows:

$$U(x,y)|_{x=0} = V_{bi} - V(y) \tag{3.24}$$

$$U(x,y)|_{x=L} = V_{bi} + V_{DS} - V(y) \tag{3.25}$$

$$U(x,y)|_{y=0} = 0 \tag{3.26}$$

$$\left.\frac{\partial U(x,y)}{\partial y}\right|_{y=t_{si}} = -\frac{\varepsilon_{ox}}{\varepsilon_{si}}\frac{U(x,y)|_{y=t_{si}}}{t_{ox}} \tag{3.27}$$

Using the typical separation of variables method for Eq. (3.23) gives a solution in the form of Fourier series [19]. The resulting expression for $\Phi(x, y)$ is

$$U(x,y)=\sum_{n=1}^{\infty}\left[A_n\sinh(k_nx)+B_n\sinh(k_n(L-x))\right]\frac{\sin(k_ny)}{\sinh(k_nL)} \tag{3.28}$$

Hence, the complete solution for $\Phi(x, y)$ can be express as:

$$\begin{aligned}\Phi(x,y)&=-\frac{qN_D}{2\varepsilon_{si}}y^2+cy+d\\&\quad+\sum_{n=1}^{\infty}\left[A_n\sinh(k_nx)+B_n\sinh(k_n(L-x))\right]\frac{\sin(k_ny)}{\sinh(k_nL)}\end{aligned} \tag{3.29}$$

In the above equations, the technology-dependent factor k_n is introduced as "natural length scale" to describe the potential distribution $\Phi(x, y)$ of the whole structure [20]. The value of k_n depends on oxide and silicon film thicknesses. Now, we need to calculate the Fourier series coefficients of A_n, B_n, and also k_n. Replacing Eq. (3.28) in the boundary condition (3.27), we have

$$\begin{aligned}&\sum_{n=1}^{\infty}\left[A_n\sinh(k_nx)+B_n\sinh(k_n(L-x))\right]\frac{k_n\cos(k_nt_{si})}{\sinh(k_nL)}\\&=-\frac{\varepsilon_{ox}}{\varepsilon_{si}}\frac{1}{t_{ox}}\sum_{n=1}^{\infty}\left[A_n\sinh(k_nx)+B_n\sinh(k_n(L-x))\right]\frac{\sin(k_nt_{si})}{\sinh(k_nL)}\end{aligned} \tag{3.30}$$

From the above equation, the factor k_n is defined as the solution of the equation:

$$\coth(k_nt_{si})=-\frac{c_{ox}}{c_{si}}\frac{1}{k_nt_{si}} \tag{3.31}$$

The factor k_n can be found by applying numerical analysis like bisection or Newton–Raphson method. It should be noted that the negative values of k_n is not the eigenvalues of (3.28) or integer. The next step is the calculation of the Fourier series coefficients A_n and B_n. Using the boundary conditions (3.24) and (3.25), we get

$$U(0,y)=\sum_{n=1}^{\infty}B_n\sin(k_ny)=V_{bi}-V(y) \tag{3.32}$$

$$U(L,y)=\sum_{n=1}^{\infty}A_n\sin(k_ny)=V_{bi}+V_{DS}-V(y) \tag{3.33}$$

To calculate A_n and B_n, we multiply both Eqs. (3.34) and (3.35) by $\sin(k_n y)$ and integrate them from 0 to t_{si} and use the orthogonality of $\sin(k_n y)$:

$$\int_0^{t_{si}} U(0,y)\sin(k_m y)dy = \int_0^{t_{si}} \sum_{n=1}^{\infty} B_m \sin(k_n y)\sin(k_m y)dy \tag{3.34}$$

$$\int_0^{t_{si}} U(L,y)\sin(k_m y)dy = \int_0^{t_{si}} \sum_{n=1}^{\infty} A_m \sin(k_n y)\sin(k_m y)dy \tag{3.35}$$

Due to the property of orthogonality of eigenfunction (3.23), we have

$$\int_0^{t_{si}} \sin(k_n y)\sin(k_m y)dy = \begin{cases} 0 & if \quad m \neq n \\ [2k_n t_{si} - \sin(2k_n t_{si})]/4k_n & if \quad m = n \end{cases} \tag{3.36}$$

Due to the independence of summation and integration functions, we can exchange them on the right-hand side of the equations and set it to zero for $m \neq n$. Thus, we can rewrite the equations as below:

$$\int_0^{t_{si}} [V_{bi} - V(y)]\sin(k_n y)dy = B_n \int_0^{t_{si}} \sin^2(k_n y)dy \tag{3.37}$$

$$\int_0^{t_{si}} [V_{bi} + V_{DS} - V(y)]\sin(k_n y)dy = A_n \int_0^{t_{si}} \sin^2(k_n y)dy \tag{3.38}$$

So, we have

$$A_n = \frac{\int_0^{t_{si}} [V_{bi} - V(y)]\sin(k_n y)dy}{\int_0^{t_{si}} \sin^2(k_n y)dy} \tag{3.39}$$

$$B_n = \frac{\int_0^{t_{si}} [V_{bi} + V_{DS} - V(y)]\sin(k_n y)dy}{\int_0^{t_{si}} \sin^2(k_n y)dy} \tag{3.40}$$

Using (3.24) and (3.25) concerning (3.17) and solving the integrals, we obtain

$$A_n = \frac{ck_n f_n + (V_{bi} + V_{DS} - d)k_n^2 \cdot g_n - h_n(qN_D/2\varepsilon_{si})}{k_n^3 \alpha_n} \tag{3.41}$$

$$B_n = \frac{ck_n f_n + (V_{bi} - d)k_n^2 \cdot g_n - h_n(qN_D/2\varepsilon_{si})}{k_n^3 \alpha_n} \tag{3.42}$$

where:

$$\alpha_n = [2k_n t_{si} - \sin(2k_n t_{si})]/4k_n \tag{3.43}$$

$$f_n = k_n t_{si} \cos(k_n t_{si}) - \sin(k_n t_{si}) \tag{3.44}$$

$$g_n = 1 - \cos(k_n t_{si}) \tag{3.45}$$

$$h_n = k_n^2 t_{si}^2 \cos(k_n t_{si}) + 2k_n t_{si} \sin(k_n t_{si}) + 2\cos(k_n t_{si}) - 2 \tag{3.46}$$

Therefore, we can express the solution $\varphi(x, y)$ from Eq. (3.29) by replacing c and d from (3.18) to (3.19) and also A_n and B_n from (3.43) to (3.44).

3.3.2 *Definition of Threshold Voltage*

The threshold voltage is a vital parameter for device modeling and characterization which plays an essential role in circuit design. This section covers the subject of extracting the threshold voltage of the device under study. There are several methods for obtaining an expression for threshold voltage [21]. Here, we use the technique which relates to the occurrence of subthreshold leakage. The subthreshold leakage current often flows at the place of the minimum bottom potential. From this point of view, the minimum bottom potential can be used as a criterion to find an expression for the threshold voltage of the short-channel MESFETs. In short-channel devices, the subthreshold current flows at the location of minimum bottom potential. Therefore, the threshold voltage can be considered as a value of gate voltage in which the minimum bottom potential becomes zero [22].

$$\Phi_{\min}(V_{GS} = V_{th}) = 0 \tag{3.47}$$

$\Phi_{\min}$ is the minimum of the channel potential. Finding the position of the minimum point along the channel ($x_{\min}$) and then evaluating $\Phi_{\min} = \Phi(x_{\min})|_{y = \text{const.}}$, the minimum channel potential is calculated [23]. To find $x_{\min}$, we equate the first derivative of Eq. (3.29) to zero:

$$\left.\frac{\partial \Phi(x, y)}{\partial x}\right|_{x = x_{\min}} = 0 \tag{3.48}$$

Differentiating (3.29) concerning x, we have

$$\sum_{n=1}^{\infty} k_n [A_n \cosh(k_n x_{\min}) - B_n \cosh(k_n (L - x_{\min}))] \frac{\sin(k_n y)}{\sinh(k_n L)} = 0 \tag{3.49}$$

After some mathematical manipulations, we obtain

$$x_{\min} = \frac{1}{2k_n} \ln \left[\frac{-A_n + B_n e^{k_n L}}{A_n - B_n e^{-k_n L}} \right] \tag{3.50}$$

With the assistance of a numerical simulation, one can use some approximations [24]. It is shown that the Fourier coefficients (A_n and B_n) decrease rapidly concerning the integer n such that the first term is nearly five times larger than the second term. Also, the hyperbolic-sine function decreases exponentially relating to the integer n. As a result, we can suppose that the first term of Fourier series is sufficient enough to represent the channel potential under the gate. Moreover, the structure of the device in the source and drain is symmetrical, and thus for $V_{DS} = 0$, we can say $A_1 = B_1$. Accordingly, the minimum surface potential can be approximated by the following equation:

$$\begin{aligned} \Phi_{\min} = & -\frac{qN_D}{2\varepsilon_{si}} t_{si}^2 + ct_{si} + d \\ & + A_1[\sinh(k_1 x_{\min}) + \sinh(k_1(L - x_{\min}))] \frac{\sin(k_1 t_{si})}{\sinh(k_1 L)} \end{aligned} \tag{3.51}$$

Substituting (3.51) into (3.47) and solving the equation for V_{GS}, the threshold voltage is found as:

$$V_{th} = \frac{\Phi_{bi} + \left(\frac{qN_D t_{si}^2}{2\varepsilon_{si}} - a_1 - a_2 t_{si}\right)(1+\gamma)}{1 + (1+\gamma)a_3} \tag{3.52}$$

The parameters a_1, a_2, and a_3 are described as:

$$\begin{aligned} a_1 = & \left[\left(\frac{\lambda_1}{\alpha_1} + \frac{a_4}{\omega_1}\right)\sinh(k_1 x_{\min}) - \frac{(a_5\Phi_{bi} - a_6)}{\alpha_1 k_1^3} \times \sinh(k_1(L_1 - x_{\min}))\right] \\ & \times \frac{\sin(k_1 t_{si})}{\sinh(k_1 L_1)} \end{aligned} \tag{3.53}$$

$$a_2 = \frac{(qN_D/c_{si})(1 - \gamma/2) + (V_{sub} - V_{fb,b})(c_{ox}/c_{si})}{(1+\gamma)} \tag{3.54}$$

$$a_3 = \left[\frac{a_7}{\omega_1}\sinh(k_1 x_{\min}) + \frac{a_5}{\alpha_1 k_1^3}\sinh(k_1(L_1 - x_{\min}))\right] \times \frac{\sin(k_1 t_{si})}{\sinh(k_1 L_1)} \tag{3.55}$$

with:

$$a_4 = b_1 \sinh(k_1(L_1 + L_2)) + b_2 \sinh(k_1 L_1) + \frac{1}{k_1^3} a_5 b_3 \sinh(k_1 L_2) \tag{3.56}$$

$$a_5 = \frac{-c_{ox} k_1 f_1}{\varepsilon_{si}(1+\gamma)} - k_1^2 g_1 \tag{3.57}$$

$$a_6 = a_2 k_1 f_1 + V_{bi} k_1^2 g_1 - h_1 (qN_D/2\varepsilon_{si}) \tag{3.58}$$

$$a_7 = \frac{1}{k_1^3} a_5 \sinh(k_1 L_2)[\sinh(k_1(L_1 + L_2) + \sinh(k_1 L_1))] \tag{3.59}$$

b_1, b_2, and b_3 in the above equations are given as:

$$b_1 = \rho_1 \sinh(k_1 L_1) + \sinh(k_1 L_2)\left[a_6/k_1^3 - \lambda_1 \cosh(k_1 L_1)\right] \tag{3.60}$$

$$b_2 = -\rho_1 \sinh(k_1 L_3)\cosh(k_1 L_2) + \frac{1}{k_1^3} a_6 \sinh(k_1 L_2) \tag{3.61}$$

$$b_3 = \Phi_{bi}\left[\sinh(k_1(L_1 + L_2) - \sinh(k_1 L_1))\right] \tag{3.62}$$

3.3.3 Definition of Subthreshold Current and Subthreshold Swing

The subthreshold behavior is one of the most critical parameters for studying the operation of the SOI MESFETs. Chang reported that the current in the subthreshold regime of MESFET could be described in the exponential form as [25]:

$$\mathrm{I}_D^{\mathrm{sub}} = I_{s0} exp\left[\left(\frac{q}{kTN_s}\right)(1 - \Gamma_1 V_{DS})(V_{GS} - V_t + \Gamma_2 V_{DS})\right], \quad V_{GS} \le V_t \tag{3.63}$$

I_{s0} is current for low drain voltage at threshold condition, q is electron charge, k is Boltzmann's constant, and T is the absolute temperature in degree Kelvin. Also, Γ_1 and Γ_2 are two coefficients for Schottky barrier height lowering. Parameter Γ_1 represents the variation in the subthreshold slope with drain voltage, and Γ_2 describes the effect of the drain bias on the threshold voltage. Also, parameter N_s indicates the efficiency of coupling between the gate and minimum channel potential which can be described as [15]:

$$N_s = \left\{ \left.\frac{\partial \Phi_{\min}}{\partial V_{GS}}\right|_{\substack{V_{GS} = V_t \\ V_{DS} = 0}} \right\}^{-1} \tag{3.64}$$

Substituting $\Phi_{\min}$ from (3.51) into (3.57), we obtain

$$N_s = \left\{\frac{1}{(1+\gamma)} - \frac{(\gamma/t_{si})f_1 + k_1 g_1(1+\gamma)}{k_1^2\alpha_1(1+\gamma)}\left[\sinh(k_1 x_{\min}) + \sinh(k_1(L - x_{\min}))\right].\frac{\sin(k_1 t_{si})}{\sinh(k_1 L)}\right\}^{-1} \tag{3.65}$$

To calculate Γ_1 and Γ_2, we can write [15]:

$$\Gamma_1 = -N_s \frac{\partial^2 \Phi_{\min}(x, t_{si})}{\partial V_{GS}\partial V_{DS}}\Bigg|_{\substack{V_{GS} = V_t \\ V_{DS} = 0}} = 0 \tag{3.66}$$

and

$$\Gamma_2 = N_s \frac{\partial \Phi_{\min}(x, t_{si})}{\partial V_{DS}}\Bigg|_{\substack{V_{GS} = V_t \\ V_{DS} = 0}}$$

$$= g_1(1+\gamma)\left\{1 - \frac{(\gamma/t_{si})f_1 + k_1 g_1(1+\gamma)}{k_1}\right\}^{-1} \tag{3.67}$$

Subthreshold swing is another critical factor in FET devices. Degradation of the subthreshold performance increases the off-current level and standby power dissipation and lessens noise immunity. Such characteristics become essential for low-power portable electronics [26]. The concept of the subthreshold swing is the amount of gate voltage needed for one-decade alteration of the drain current and is defined as the inverse of the slope of the log I_D against V_{GS} curve in the subthreshold region [27] as follows:

$$SS = \left\{\frac{\partial \log I_D^{sub}}{\partial V_{GS}}\right\}^{-1} \tag{3.68}$$

By ignoring the reverse conduction current of the gate–drain junction, Eq. (3.47) can be described as:

$$SS = (\ln 10)\frac{kT}{q}N_s \tag{3.69}$$

Substituting Ns from (3.51) into (3.61), we obtain

$$SS = (\ln 10) \times \frac{kT}{q}\left\{\frac{1}{(1+\gamma)} - \frac{(\gamma/t_{si})f_1 + k_1 g_1(1+\gamma)}{k_1^2\alpha_1(1+\gamma)}\left[\sinh(k_1 x_{\min}) + \sinh(k_1(L - x_{\min}))\right].\frac{\sin(k_1 t_{si})}{\sinh(k_1 L)}\right\}^{-1} \tag{3.70}$$

3.4 Conclusion

Analytical modeling of SOI MESFET is generally obtained by solving Poisson's equation to derive an expression for channel potential. Superposition method is an accurate technique for solving Poisson's equation, in which the solution of the 2-D Poisson's equation is represented as the sum of the 1-D Poisson's equation and a 2-D Laplace's equation solutions. The general solution of Laplace equation is described in the form of Fourier series. To analytically model the device, the Fourier coefficients must be represented by device parameters. In the previous two models, the authors could not find a methodology to solve the summation equation of Fourier series. Therefore, in their solution, they used invalid assumption which led them to inaccurate results.

In this chapter, we employed a new methodology to find the coefficients of Fourier series. We applied the boundary conditions and multiplied the resultant equation by $\mathrm{sim}(k_m y)$. Then, we applied integral from 0 to t_{si} and used the property of orthogonality and evaluated the integrals to find the Fourier series coefficients. The advantages of the presented models are:

- The model is more accurate than two others.
- The model is correct for all values of device parameters.
- The parameter k_n depends on device parameters.
- For all values of n, the parameter k_n has a defined value.

References

1. K.P. MacWilliams, J.D. Plummer, in *Electron Devices Meeting, 1986 International*. Complementary silicon MESFETs for VLSI (IEEE, 1986), pp. 403–406
2. P.E. Cottrell, R.R. Troutman, T.H. Ning, Hot-electron emission in n-channel IGFETs. IEEE J. Solid State Circuits **14**, 442–455 (1979)
3. E. Rimini, *Ion Implantation: Basics to Device Fabrication* (Springer, Berlin, 2013)
4. A. Chaudhry, *Fundamentals of Nanoscaled Field Effect Transistors* (Springer, Berlin, 2013)
5. K.P. MacWilliams, J.D. Plummer, Device physics and technology of complementary silicon MESFET's for VLSI applications. IEEE Trans. Electron Devices **38**, 2619–2631 (1991)
6. J.D. Marshall, J.D. Meindl, A sub- and near-threshold current model for silicon MESFETs. IEEE Trans. Electron Devices **35**, 388–390 (1988)
7. J. Nulman, J.V. Faricelli, J.P. Krusius, J. Frey, Fabrication and analysis of 1/2μm silicon logic MESFET's. IEEE Trans. Electron Devices **30**, 1395–1401 (1983)
8. A.A. Orouji, Z. Ramezani, S.M. Sheikholeslami, A novel SOI-MESFET structure with double protruded region for RF and high voltage applications. Mater. Sci. Semicond. Process. **30**, 545–553 (2015)
9. S. Cristoloveanu, S. Li, *Electrical Characterization of Silicon-on-Insulator Materials and Devices* (Springer, Berlin, 2013)
10. T. Houston, C. Everett, H. Darley, G. Taylor, in *Solid-State Circuits Conference. Digest of Technical Papers. 1979 IEEE International*. Silicon MESFET circuit performance for VLSI (IEEE, 1979), pp. 80–81

11. P.A. Tove, K. Bohlin, F. Masszi, H. Norde, J. Nylander, J. Tiren, U. Magnusson, Complementary Si MESFET concept using silicon-on-sapphire technology. Electron Device Lett. IEEE **9**, 47–49 (1988)
12. K.K. Young, Short-channel effect in fully depleted SOI MOSFETs. IEEE Trans. Electron Devices **36**, 399–402 (1989)
13. P. Pandey, B.B. Pal, S. Jit, A new 2-D model for the potential distribution and threshold voltage of fully depleted short-channel Si-SOI MESFETs. IEEE Trans. Electron Devices **51**, 246–254 (2004)
14. M.A. Imam, M.A. Osman, A.A. Osman, Threshold voltage model for deep-submicron fully depleted SOI MOSFETs with back gate substrate induced surface potential effects. Microelectron. Reliab. **39**, 487–495 (1999)
15. J.D. Marshall, J.D. Meindl, An analytical two-dimensional model for silicon MESFETs. IEEE Trans. Electron Devices **35**, 373–383 (1988)
16. T.K. Chiang, Y.H. Wang, M.P. Houng, Modeling of threshold voltage and subthreshold swing of short-channel SOI MESFET's. Solid State Electron. **43**, 123–129 (1999)
17. K. Suzuki, S. Pidin, Short-channel single-gate SOI MOSFET model. IEEE Trans. Electron Devices **50**, 1297–1305 (2003)
18. S.K. Lahiri, A. DasGupta, I. Manna, M.K. Das, A quasi-3D analytical threshold voltage model of small geometry MOSFETs. Solid State Electron. **35**, 1721–1727 (1992)
19. K.F. Riley, M.P. Hobson, S.J. Bence, *Mathematical Methods for Physics and Engineering: A Comprehensive Guide* (Cambridge University Press, Cambridge, 2006)
20. R.H. Yan, A. Ourmazd, K.F. Lee, Scaling the Si MOSFET: From bulk to SOI to bulk. IEEE Trans. Electron Devices **39**, 1704–1710 (1992)
21. A. Ortiz-Conde, F.J. García-Sánchez, J. Muci, A. Terán Barrios, J.J. Liou, C.-S. Ho, Revisiting MOSFET threshold voltage extraction methods. Microelectron. Reliab. **53**, 90–104 (2013)
22. T.K. Chiang, in *9th International Conference on Solid-State and Integrated-Circuit Technology, 2008. ICSICT 2008*. The new analytical subthreshold behavior model for dual material gate (DMG) SOI MESFET (2008), pp. 288–289
23. N. Lakhdar, F. Djeffal, A two-dimensional analytical model of subthreshold behavior to study the scaling capability of deep submicron double-gate GaN-MESFETs. J. Comput. Electron. **10**, 382–387 (2011)
24. S.P. Chin, P.D. Ching-Yuan Wu, A new two-dimensional model for the potential distribution of short gate-length MESFET's and its applications. IEEE Trans. Electron Devices **39**, 1928–1937 (1992)
25. C.T.M. Chang, T. Vrotsos, M.T. Frizzell, R. Carroll, A subthreshold current model for GaAs MESFET's. IEEE Electron Device Lett. **8**, 69–72 (1987)
26. A. Godoy, J.A. López-Villanueva, J.A. Jiménez-Tejada, A. Palma, F. Gámiz, A simple subthreshold swing model for short channel MOSFETs. Solid State Electron. **45**, 391–397 (2001)
27. B.-G. Park, in *Nano Devices and Circuit Techniques for Low-Energy Applications and Energy Harvesting*, ed. by C.-M. Kyung (Springer, Dordrecht, 2016), pp. 3–31

Chapter 4
Design and Modeling of Triple-Material Gate SOI MESFET

4.1 Introduction

Gate material engineering is another technique which provides simultaneous carrier transport enhancement and suppression of short-channel effects. It is obtained by robust control of the gate material work-function and also the length of the beside merged gate materials [1]. In this regard, dual-material gate devices have been emerged using innovation in the gate material [2]. The most important feature of these devices is enhancing the carrier transport efficiency. The transport efficiency of the gate is related to the average velocity of electrons traveling through the conductive channel, which is governed by the electric field along the channel. In an MOS transistor, electrons are entered into the channel with a low primary velocity and gradually accelerate when they move toward the drain. The drift velocity of electrons becomes maximum when they reach near the drain. The movement of electrons is very fast in the area near the drain but comparatively slow in the area nearby the source. This means that the relatively slow speed of electron transit nearby the source region has lessened the speed of the MOS devices.

The DMG structure modifies the surface potential profile and the electric field pattern lengthwise the channel so that the transport efficiency of gate increases [3]. In this way, a step potential is induced at the boundary between two dissimilar gate materials and form a peak electric field in the conductive channel which improves the speed of carrier transport and increases driving capability of the device. Regarding the screen effect from the material with lower work-function, the high electric field nearby the drain are effectively reduced, which, in turn, reduces the hot carrier effect (HCE) and the current leakage of the substrate [4]. On the other hand, the discontinuity of electric field at the border of three gate materials decreases the steepness in the electric field profile, and the whole of the channel field becomes flattened which results in a more average velocity of electrons which are entered into the channel from the source terminal. In comparison to conventional MOS devices, the DMG devices offer better current-driving capability. However, the gate transport

I. S. Amiri et al., *Device Physics, Modeling, Technology, and Analysis for Silicon MESFET*, https://doi.org/10.1007/978-3-030-04513-5_4

efficiency of the device can be more improved by incorporating triple-material gate instead of the dual-material gate. Dual-metal gate induces one step potential, whereas triple-metal gate (TMG) devices offer two step-potential effects [5].

To combine the advantages of silicon SOI MESFET and triple-material gate, the new structure called triple-material gate silicon SOI MESFET is introduced. By the exact solution of 2-D Poisson's equation with suitable boundary conditions, an analytical model for the surface potential, threshold voltage, and short-channel effects is developed. The influence of device parameters and applied biases are studied and investigated.

4.2 Description of the Device Structure

Figure 4.1 demonstrates the 2D schematic view of a tri-material gate MESFET with L_1, L_2, and L_3 as the length of each gate metals. In this device, the gate material consists of three dissimilar metals with different work-functions which are connected alongside together. Thus, the surface potential in the channel is distributed as Φ_1, Φ_2, and Φ_3 in the regions under metals M_1, M_2, and M_3, respectively. Region one is defined as $0 \leq x \leq L_1$ and $0 \leq y \leq t_{si}$, similarly region two as $L_1 \leq x \leq L_1 + L_2$ and $0 \leq y \leq t_{si}$, and region three as $L_1 + L_2 \leq x \leq L_1 + L_2 + L_3$ and $0 \leq y \leq t_{si}$. Likewise, the total channel length is defined as $L = L_1 + L_2 + L_3$.

The work-functions are chosen as follows: the metal with highest work-function (M_1) is placed adjacent to the source, the metal with lowest work-function (M_3) is positioned adjacent to the drain, and the metal with transitional work-function (M_2) is placed between two other metals.

This structure induces two discontinuities in surface potential profile which screens the influence of drain in the channel. Here, the first gate metal (M_1) is called control gate, the middle gate metal (M_2) is called the first screen gate, and the last gate metal on the drain side (M_3) is called the second screen gate.

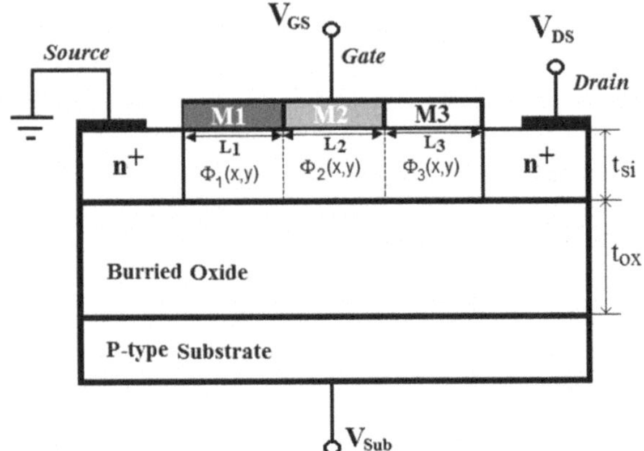

Fig. 4.1 The schematic cross-sectional view of a tri-material gate silicon SOI MESFET

4.3 Potential Distribution

To investigate and explain the proposed MESFET structure, we need to analyze device performance by developing an analytical model to derive an expression for the potential distribution in two dimensions. Supposing that the impurity density in the active channel is uniform and ignoring the effect of mobile carriers and fixed trapped charge on the electrostatic of the channel, the Poisson's equations of potential for three regions before the onset of high inversion are described as:

$$\frac{\partial^2 \Phi_1(x,y)}{\partial x^2} + \frac{\partial^2 \Phi_1(x,y)}{\partial y^2} = -\frac{qN_D}{\varepsilon_{\text{si}}} \tag{4.1}$$

$$\frac{\partial^2 \Phi_2(x,y)}{\partial x^2} + \frac{\partial^2 \Phi_2(x,y)}{\partial y^2} = -\frac{qN_D}{\varepsilon_{\text{si}}} \tag{4.2}$$

$$\frac{\partial^2 \Phi_3(x,y)}{\partial x^2} + \frac{\partial^2 \Phi_3(x,y)}{\partial y^2} = -\frac{qN_D}{\varepsilon_{\text{si}}} \tag{4.3}$$

4.3.1 Mathematical Approach

The solution of the above equations can be attained by using the superposition method. According to this, we decompose the 2-D Poisson's equation into the 1-D Poisson's equation and 2-D Laplace equation. It leads to:

$$\Phi_i(x,y) = V_i(y) + U_i(x,y) \tag{4.4}$$

Therefore, $V_i(y)$ is the solution of 1D Poisson's equation and is described as:

$$\frac{\mathrm{d}^2 V_i(y)}{\partial y^2} = -\frac{qN_D}{\varepsilon_{\text{si}}} \tag{4.5}$$

Likewise, $U_i(x, y)$ is the solution of 2D Laplace's equation and is represented as:

$$\frac{\partial^2 U_i(x,y)}{\partial x^2} + \frac{\partial^2 U_i(x,y)}{\partial y^2} = 0 \tag{4.6}$$

Index i ($i = 1, 2, 3$) indicates the area under each gate materials (M_1, M_2, and M_3).

4.3.2 Boundary Conditions

Figure 4.2 shows the expanded view of the rectangle channel area with boundaries of three regions for the solution of 2-D Poisson's equation.

To solve Eqs. (4.59)–(4.61), we require a proper set of boundary conditions as follows.

The Left and Right Boundary Conditions

The left and right boundaries are the junction between the channel and source–drain areas, respectively. Similar to single-gate SOI MESFET, the potential at the source and drain terminals are described as:

$$\Phi_1(x,y) = V_{bi} \tag{4.7}$$

$$\Phi_3(L_1 + L_2 + L_3, y) = V_{bi} + V_{DS} \tag{4.8}$$

where V_{bi} is the built-in potential of the source–channel and drain–channel junctions and V_{DS} is applied drain-to-source voltage.

The Top Boundary Condition

The top boundary is the border between three gate materials and the silicon film. The potential at the top of the silicon body for each material is defined as:

$$\Phi_1(x,0) = V_{GS} - \Phi_{bi1} \tag{4.9}$$

$$\Phi_2(x,0) = V_{GS} - \Phi_{bi2} \tag{4.10}$$

$$\Phi_3(x,0) = V_{GS} - \Phi_{bi3} \tag{4.11}$$

Here, Φ_{bi1}, Φ_{bi2}, and Φ_{bi3} are the built-in potentials of the Schottky barriers formed by three gate metals and silicon film and V_{GS} is the applied gate-to-source voltage. Since three different materials with different work-functions are used for

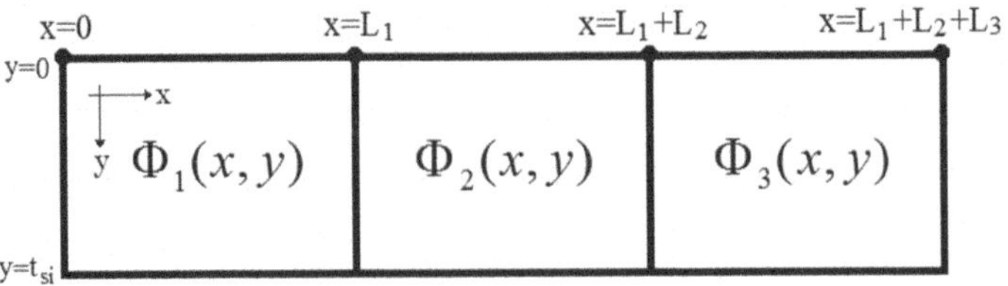

Fig. 4.2 The expanded view of the rectangle channel area

gate metal, the built-in potential for each gate material would be different, as it depends on the metal work-functions Φ_{M1}, Φ_{M2}, and Φ_{M3}.

The Boundaries Between Two Channel Areas Under Two Dissimilar Materials

The potential at the interface of three dissimilar materials is continuous. So, we can write

$$\Phi_1(L_1, y) = \Phi_2(L_1, y) \tag{4.12}$$

$$\Phi_2(L_1 + L_2, y) = \Phi_3(L_1 + L_2, y) \tag{4.13}$$

Also, the electric field at the interface of three dissimilar materials is continuous. It yields

$$\frac{d\Phi_1(L_1, y)}{dx} = \frac{d\Phi_2(L_1, y)}{dx} \tag{4.14}$$

$$\frac{d\Phi_2(L_1 + L_2, y)}{dx} = \frac{d\Phi_3(L_1 + L_2, y)}{dx} \tag{4.15}$$

Bottom Boundary Condition

The bottom boundary is the border between the silicon film and buried oxide. Electric flux at the interface of silicon–back oxide for each gate material is continuous. It leads us to:

$$\frac{d\Phi_1(x, t_{si})}{\partial y} = \frac{\varepsilon_{ox}}{\varepsilon_{si}} \frac{V_{sub} - V_{fb}^{b} - \Phi_1(x, t_{si})}{t_{ox}} \tag{4.16}$$

$$\frac{d\Phi_2(x, t_{si})}{\partial y} = \frac{\varepsilon_{ox}}{\varepsilon_{si}} \frac{V_{sub} - V_{fb}^{b} - \Phi_2(x, t_{si})}{t_{ox}} \tag{4.17}$$

$$\frac{d\Phi_3(x, t_{si})}{\partial y} = \frac{\varepsilon_{ox}}{\varepsilon_{si}} \frac{V_{sub} - V_{fb}^{b} - \Phi_3(x, t_{si})}{t_{ox}} \tag{4.18}$$

where V_{sub} is the substrate voltage, ε_{ox} is the permittivity of the buried oxide, and t_{ox} is the thickness of buried oxide layer. Also, V_{fb}^{b} is the flat-band voltage of the substrate which is given as:

$$V_{\text{fb}}^{b} = \frac{kT}{q} \ln\left(\frac{N_{\text{sub}}}{N_D}\right) \tag{4.19}$$

4.3.3 Solution of $V_i(y)$

By two times integrating on Eq. (4.5) concerning y, we have

$$\frac{\mathrm{d}V_i(y)}{\mathrm{d}y} = -\frac{qN_D}{\varepsilon_{\text{si}}}y + c_i \tag{4.20}$$

and

$$V_i(y) = -\frac{qN_D}{2\varepsilon_{\text{si}}}y^2 + c_i y + d_i \tag{4.21}$$

To find the integration constants c_i and d_i, we make use of the following boundary conditions:

$$V_1(0) = V_{\text{GS}} - \Phi_{\text{bi1}} \tag{4.22}$$

$$V_2(0) = V_{\text{GS}} - \Phi_{\text{bi2}} \tag{4.23}$$

$$V_3(0) = V_{\text{GS}} - \Phi_{\text{bi3}} \tag{4.24}$$

and

$$\frac{\mathrm{d}V_1(t_{\text{si}})}{\partial y} = \frac{\varepsilon_{\text{ox}}}{\varepsilon_{\text{si}}} \frac{\left(V_{\text{sub}} - V_{\text{fb}}^{b}\right) - V_1(t_{\text{si}})}{t_{\text{ox}}} \tag{4.25}$$

$$\frac{\mathrm{d}V_2(t_{\text{si}})}{\partial y} = \frac{\varepsilon_{\text{ox}}}{\varepsilon_{\text{si}}} \frac{\left(V_{\text{sub}} - V_{\text{fb}}^{b}\right) - V_2(t_{\text{si}})}{t_{\text{ox}}} \tag{4.26}$$

$$\frac{\mathrm{d}V_3(t_{\text{si}})}{\partial y} = \frac{\varepsilon_{\text{ox}}}{\varepsilon_{\text{si}}} \frac{\left(V_{\text{sub}} - V_{\text{fb}}^{b}\right) - V_3(t_{\text{si}})}{t_{\text{ox}}} \tag{4.27}$$

Using boundary conditions (4.25)–(4.27) into Eq. (4.21), the solution for d_1, d_2, and d_3 is obtained as:

$$d_1 = V_{\text{GS}} - \Phi_{\text{bi1}} \tag{4.28}$$

$$d_2 = V_{\text{GS}} - \Phi_{\text{bi2}} \tag{4.29}$$

$$d_3 = V_{\mathrm{GS}} - \Phi_{\mathrm{bi3}} \tag{4.30}$$

Substituting Eqs. (4.20) and (4.21) with $i = 1$ into (4.25) and solving the equation for c_1, we have

$$c_1 = \frac{(\gamma/t_{\mathrm{si}})\left(V_{\mathrm{sub}} - V_{\mathrm{fb}}^{b} - d_1\right) + (qN_D/c_{\mathrm{si}})(1 + \gamma/2)}{1 + \gamma} \tag{4.31}$$

where $\gamma = c_{\mathrm{ox}}/c_{\mathrm{si}}$.

Using same way for c_2 and c_3, we obtain

$$c_2 = \frac{(\gamma/t_{\mathrm{si}})\left(V_{\mathrm{sub}} - V_{\mathrm{fb}}^{b} - d_2\right) + (qN_D/c_{\mathrm{si}})(1 + \gamma/2)}{1 + \gamma} \tag{4.32}$$

$$c_3 = \frac{(\gamma/t_{\mathrm{si}})\left(V_{\mathrm{sub}} - V_{\mathrm{fb}}^{b} - d_3\right) + (qN_D/c_{\mathrm{si}})(1 + \gamma/2)}{1 + \gamma} \tag{4.33}$$

4.3.4 *Solution of* U$_i$(x, y)

By using the separation of variables method together with the given boundary conditions, one can obtain the following resultant solution of 2-D Laplace equation:

$$U_i(x, y) = \sum_{n=1}^{\infty} \left[A_n \sinh(k_n x) + B_n \sinh(k_n(L_1 - x))\right] \frac{\sin(k_n y)}{\sinh(k_n L_1)} \tag{4.34}$$

$$U_2(x, y) = \sum_{n=1}^{\infty} \left[C_n \sinh(k_n(x - L_1)) + D_n \sinh(k_n(L_1 + L_2 - x))\right] \frac{\sin(k_n y)}{\sinh(k_n L_2)} \tag{4.35}$$

$$U_3(x, y) = \sum_{n=1}^{\infty} \left[E_n \sinh(k_n(x - L_1 - L_2)) + F_n \sinh(k_n(L_1 + L_2 + L_3 - x))\right] \times \frac{\sin(k_n y)}{\sinh(k_n L_3)} \tag{4.36}$$

In the next step, the Fourier series coefficients A_n, B_n, C_n, D_n, E_n, and F_n and also eigenvalue k_n must be calculated. Regarding to $\Phi_i(x, y) = V_i(y) + U_i(x, y)$, we rewrite the boundary conditions (4.12)–(4.18) for $U_i(x, y)$ as follows:

$$U_1(L_1, y) + V_1(y) = U_2(L_1, y) + V_2(y) \tag{4.37}$$

$$U_2(L_1 + L_2, y) + V_2(y) = U_3(L_1 + L_2, y) + V_3(y) \tag{4.38}$$

$$\frac{dU_1(L_1, y)}{\partial x} = \frac{dU_2(L_1, y)}{\partial x} \tag{4.39}$$

$$\frac{dU_2(L_1 + L_2, y)}{\partial x} = \frac{dU_3(L_1 + L_2, y)}{\partial x} \tag{4.40}$$

$$\frac{dU_1(x, t_{si})}{\partial y} = \frac{\varepsilon_{ox}}{\varepsilon_{si}} \frac{-U_1(x, t_{si})}{t_{ox}} \tag{4.41}$$

$$\frac{dU_2(x, t_{si})}{\partial y} = \frac{\varepsilon_{ox}}{\varepsilon_{si}} \frac{-U_2(x, t_{si})}{t_{ox}} \tag{4.42}$$

$$\frac{dU_3(x, t_{si})}{\partial y} = \frac{\varepsilon_{ox}}{\varepsilon_{si}} \frac{-U_3(x, t_{si})}{t_{ox}} \tag{4.43}$$

Replacing boundary conditions (4.41)–(4.43) into Eqs. (4.34)–(4.36) gives us the eigenvalue k_n for all three regions of the channel as follows:

$$\coth(k_n t_{si}) = -\frac{c_{ox}}{c_{si}} \frac{1}{k_n t_{si}} \tag{4.44}$$

To calculate B_n, we apply boundary condition (4.6) on Eqs. (4.34) and (4.21), so we obtain

$$\sum_{n=1}^{\infty} B_n \sin(k_n y) = V_{bi} \tag{4.45}$$

Now, we multiply Eq. (4.45) by $\sin(k_m y)$ and integrate from 0 to t_{si}:

$$\int_0^{t_{si}} \sum_{n=1}^{\infty} B_n \sin(k_n y) \sin(k_m y) dy = \int_0^{t_{si}} [V_{bi} - V_1(y)] \sin(k_m y) dy \tag{4.46}$$

Now, we apply the property of orthogonality of eigenfunctions (4.4), so we can write

$$\int_0^{t_{si}} \sin(k_n y) \sin(k_m y) dy = \begin{cases} 0, & m \neq n \\ [2k_n t_{si} - \sin(2k_n t_{si})]/4k_n, & m = n \end{cases} \tag{4.47}$$

As both functions of integration and summation are independent and linear, we rewrite the Eq. (4.46) for B_n when $m = n$ as:

$$B_n = \frac{\int_0^{t_{\text{si}}} [V_{\text{bi}} - V_1(y)] \sin(k_n y) dy}{\int_0^{t_{\text{si}}} \sin^2(k_n y) dy} = \frac{\beta_n}{\alpha_n} \tag{4.48}$$

After evaluating the integrals and using some mathematical simplification, we get

$$\alpha_n = [2k_n t_{\text{si}} - \sin(2k_n t_{\text{si}})]/4k_n \tag{4.49}$$

$$\beta_n = \frac{1}{k_n^3}\left[c_1 k_n f_n + (V_{\text{bi}} - d_1) \cdot k_n^2 g_n - h_n(qN_D/2\varepsilon_{\text{si}})\right] \tag{4.50}$$

where:

$$f_n = k_n t_{\text{si}} \cos(k_n t_{\text{si}}) - \sin(k_n t_{\text{si}}) \tag{4.51}$$

$$g_n = 1 - \cos(k_n t_{\text{si}}) \tag{4.52}$$

$$h_n = (k_n^2 t_{\text{si}}^2 - 2)\cos(k_n t_{\text{si}}) - 2k_n t_{\text{si}} \sin(k_n t_{\text{si}}) + 2 \tag{4.53}$$

To calculate E_n, we introduce boundary condition (4.8) on Eq. (4.36). So, we have

$$V_3(y) + \sum_{n=1}^{\infty} E_n \sin(k_n y) = V_{\text{bi}} + V_{\text{DS}} \tag{4.54}$$

Using the same approach used for (4.45), we obtain

$$B_n = \frac{\gamma_n}{\alpha_n} \tag{4.55}$$

with:

$$\gamma_n = \frac{1}{k_n^3}\left[c_3 k_n f_n + (V_{\text{bi}} + V_{\text{DS}} - d_3) \cdot k_n^2 g_n - h_n(qN_D/2\varepsilon_{\text{si}})\right] \tag{4.56}$$

Substituting Eqs. (4.34)–(4.36) into (4.37)–(4.40), we obtain

$$\sum_{n=1}^{\infty} (A_n - D_n) \sin(k_n y) = (c_2 - c_1)y + (d_2 - d_1) \tag{4.57}$$

$$\sum_{n=1}^{\infty}(C_n - F_n)\sin(k_n y) = (c_3 - c_2)y + (d_3 - d_2) \tag{4.58}$$

$$\sum_{n=1}^{\infty}[A_n\cosh(k_n L_1) - B_n]\cdot\frac{1}{\sinh(k_n L_1)} = \sum_{n=1}^{\infty}[C_n - D_n\cosh(k_n L_2)]\cdot\frac{1}{\sinh(k_n L_2)} \tag{4.59}$$

$$\sum_{n=1}^{\infty}[C_n\cosh(k_n L_2) - D_n]\cdot\frac{1}{\sinh(k_n L_2)} = \sum_{n=1}^{\infty}[E_n - F_n\cosh(k_n L_3)]\cdot\frac{1}{\sinh(k_n L_3)} \tag{4.60}$$

Now, we multiply Eqs. (4.57) and (4.58) by $\sin(k_m y)$ and integrate from 0 to t_{si}. Also, we can remove summation from Eqs. (4.59) and (4.60). So, we have

$$\sum_{n=1}^{\infty}(A_n - D_n)\sin(k_n y) = (c_2 - c_1)y + (d_2 - d_1) \tag{4.61}$$

$$\sum_{n=1}^{\infty}(C_n - F_n)\sin(k_n y) = (c_3 - c_2)y + (d_3 - d_2) \tag{4.62}$$

$$\sum_{n=1}^{\infty}[A_n\cosh(k_n L_1) - B_n]\cdot\frac{1}{\sinh(k_n L_1)} = \sum_{n=1}^{\infty}[C_n - D_n\cosh(k_n L_2)]\cdot\frac{1}{\sinh(k_n L_2)} \tag{4.63}$$

$$\sum_{n=1}^{\infty}[C_n\cosh(k_n L_2) - D_n]\cdot\frac{1}{\sinh(k_n L_2)} = \sum_{n=1}^{\infty}[E_n - F_n\cosh(k_n L_3)]\cdot\frac{1}{\sinh(k_n L_3)} \tag{4.64}$$

Now, we multiply Eqs. (4.57) and (4.58) by $\sin(k_m y)$ and integrate from 0 to t_{si}. Also, we can remove summation from Eqs. (4.59) and (4.60). So, we have

$$(A_n - D_n)\int_0^{t_{\mathrm{si}}}\sin^2(k_n y)\mathrm{d}y = \int_0^{t_{\mathrm{si}}}[(c_2 - c_1)y + (d_2 - d_1)]\sin(k_n y)\mathrm{d}y \tag{4.65}$$

$$(C_n - F_n)\int_0^{t_{si}}\sin^2(k_n y)\mathrm{d}y = \int_0^{t_{si}}[(c_3 - c_2)y + (d_3 - d_2)]\sin(k_n y)\mathrm{d}y \tag{4.66}$$

$$[A_n\cosh(k_n L_1) - B_n]\sinh(k_n L_2) = [C_n - D_n\cosh(k_n L_2)]\sinh(k_n L_1) \tag{4.67}$$

$$[C_n\cosh(k_n L_2) - D_n]\sinh(k_n L_3) = [E_n - F_n\cosh(k_n L_3)]\sinh(k_n L_2) \tag{4.68}$$

Evaluating the integrals and using some mathematical manipulation, we get

$$A_n - D_n = \frac{\lambda_n}{\alpha_n} \tag{4.69}$$

$$C_n - F_n = \frac{\rho_n}{\alpha_n} \tag{4.70}$$

$$D_n \sinh(k_n(L_1 + L_2)) - F_n \sinh(k_n L_1) = \frac{\delta_n}{\alpha_n} \tag{4.71}$$

$$-D_n \sinh(k_n L_3) + F_n \sinh(k_n(L_1 + L_2)) = \frac{\eta_n}{\alpha_n} \tag{4.72}$$

where:

$$\delta_n = \rho_n \sinh(k_n L_1) + [\beta_n - \lambda_n \cosh(k_n L_1)] \sinh(k_n L_2) \tag{4.73}$$

$$\eta_n = -\rho_n \sinh(k_n L_3) \cosh(k_n L_2) + \gamma_n \sinh(k_n L_2) \tag{4.74}$$

$$\lambda_n = [-f_n(c_2 - c_1) + g_n(d_2 - d_1)]/k_n^2 \tag{4.75}$$

$$\rho_n = [-f_n(c_3 - c_2) + g_n(d_3 - d_2)]/k_n^2 \tag{4.76}$$

Using Eqs. (4.69)–(4.72), we get a system of equations in A_n, D_n, C_n, and F_n. Solving the system of equations, we obtain

$$A_n = \frac{\lambda_n}{\alpha_n} + D_n \tag{4.77}$$

$$C_n = \frac{\rho_n}{\alpha_n} + F_n \tag{4.78}$$

$$D_n = [\delta_n \sinh(k_n(L_2 + L_3)) + \eta_n \sinh(k_n L_1)]/\omega_n \tag{4.79}$$

$$F_n = [\delta_n \sinh(k_n L_3) + \eta_n \sinh(k_n(L_1 + L_2))]/\omega_n \tag{4.80}$$

where:

$$\omega_n = \alpha_n [\sinh(L_1 + L_2) \cdot \sinh(k_n(L_2 + L_3)) - \sinh(k_n L_1) \cdot \sinh(k_n L_3)] \tag{4.81}$$

4.4 Definition of Threshold Voltage

The threshold voltage is defined as the gate voltage in which the minimum bottom potential being zero. In threshold condition, the portion of the channel with higher work-function will not conduct the channel with lower work-function. This means that the threshold behavior is mostly governed by the gate with material 1 (M_1). Regarding Eqs. (4.21) and (4.34) with $i = 1$, the channel potential under the gate material 1 can be described as:

$$\Phi_1(x,y) = -\frac{qN_D}{2\varepsilon_{si}}y^2 + c_1 y + d_1 + \sum_{n=1}^{\infty}[A_n \sinh(k_n x) + B_n \sinh(k_n(L_1 - x))]\frac{\sin(k_n y)}{\sinh(k_n L_1)} \quad (4.82)$$

Due to the rapid decay of the Fourier series coefficients A_n and B_n in Eq. (4.34), the first terms A_1 and B_1 are proper to express the entire series [4]. By setting the first derivative of Eq. (4.82) at $y =$ constant to zero, the minimum surface potential and its location can be obtained as:

$$k_1[A_1 \cosh(k_1 x_{min}) - B_1 \cosh(k_1(L_1 - x_{min}))]\frac{\sin(k_1 t_{si})}{\sinh(k_1 L_1)} = 0 \quad (4.83)$$

Solving the above equation for x_{min}, we have

$$x_{min} = \frac{1}{2k_1}\ln\left[\frac{-A_1 \exp(k_1 L_1) + B_1}{A_1 - B_1 \exp(-k_1 L_1)}\right] \quad (4.84)$$

The minimum channel potential at $y = t_{si}/8$ is found by replacing x_{min} from (4.84) into (4.82):

$$\Phi_1(\min) = -\frac{qN_D}{128\varepsilon_{si}}t_{si}^2 + \frac{1}{8}c_1 t_{si} + d_1 + [A_1 \sinh(k_1 x) + B_1 \sinh(k_1(L_1 - x))]\frac{\sin(k_1 t_{si}/8)}{\sinh(k_1 L_1)} \quad (4.85)$$

By deriving V_{GS} from (4.85) for $\Phi_{1,min} = 0$, the threshold voltage can be found as:

$$V_{th} = \frac{\Phi_{bi1} + (qN_D t_{si}^2/2\varepsilon_{si} - a_1 - a_2)(1+\gamma)}{1 + a_3(1+\gamma)} \quad (4.86)$$

with:

$$a_1 = \left[\left(\frac{\lambda_1}{a_1} + \frac{a_4}{\omega_1}\right)\sinh(k_1 x_{min}) - \frac{(a_5 \Phi_{bi1} - a_6)}{a_1 k_1^3}\sinh(k_1(L_1 - x_{min}))\right]\frac{\sin(k_1 t_{si})}{\sinh(k_1 L_1)} \quad (4.87)$$

$$a_2 = \frac{(qN_D/c_{si})(1-\gamma/2) + \left(V_{sub} - V_{fb}^b\right)(c_{ox}/\varepsilon_{si})}{1+\gamma} \tag{4.88}$$

$$a_3 = \left[\frac{a_7}{\omega_1}\sinh(k_1 x_{min}) + \frac{a_5}{a_1 k_1^3}\sinh(k_1(L_1 - x_{min}))\right]\frac{\sin(k_1 t_{si})}{\sinh(k_1 L_1)} \tag{4.89}$$

The parameters a_4, a_5, a_6, and a_7 in the above equations are expressed as:

$$a_4 = b_1\sinh(k_1(L_1+L_2)) + b_2\sinh(k_1L_1) + \frac{1}{k_1^3}a_5b_3\sinh(k_1L_2) \tag{4.90}$$

$$a_5 = \frac{-c_{ox}k_1f_1}{\varepsilon_{si}(1+\gamma)} - k_1^2g_1 \tag{4.91}$$

$$a_6 = a_2k_1f_1 + V_{bi}k_1^2g_1 - h_1(qN_D/2\varepsilon_{si}) \tag{4.92}$$

b_1, b_2, and b_3 in the above equations are given as:

$$b_1 = \rho_1\sinh(k_1L_1) + \frac{1}{k_1^3}a_6\sinh(k_1L_2) - \lambda_1\sinh(k_1L_2)\cosh(k_1L_1) \tag{4.93}$$

$$b_2 = -\rho_1\sinh(k_1L_3)\cosh(k_1L_2) + \frac{1}{k_1^3}\left(a_6 + V_{DS}k_1^2g_1\right)\sinh(k_1L_2) \tag{4.94}$$

$$b_3 = [\sinh(k_1(L_2+L_3)) - \sinh(k_1L_1)]\Phi_{bi1} \tag{4.95}$$

References

1. A. Chaudhry, *Analytical Modeling and Simulation of Short-Channel Effects in a Fully Depleted Dual-Material Gate (DMG) SOI MOSFET* (Indian Institute of Technology Delhi, Delhi, 2003)
2. W. Long, H. Ou, J.M. Kuo, K.K. Chin, Dual-material gate (DMG) field effect transistor. IEEE Trans. Electron Devices **46**, 865–870 (1999)
3. K. Goel, M. Saxena, M. Gupta, R. Gupta, Comparison of three region multiple gate nanoscale structures for reduced short channel effects and high device reliability. NSIT Nanotechnol. **3**, 816–819 (2006)
4. T.-K. Chiang, A new two-dimensional analytical subthreshold behavior model for short-channel tri-material gate-stack SOI MOSFET's. Microelectron. Reliab. **49**, 113–119 (2009)
5. A. Priya, R.A. Mishra, A two dimensional analytical modeling of surface potential in triple metal gate (TMG) fully-depleted recessed-source/drain (Re-S/D) SOI MOSFET. Superlattice Microst. **92**, 316–329 (2016)

Chapter 5
Three-Dimensional Analytical Model of the Non-Classical Three-Gate SOI MESFET

5.1 Introduction

Multiple-gate structure is an innovative idea which provides the opportunity for further miniaturization owing to its remarkable current-driving capability which is originated from the presence of several gates that offer more intensive control of the channel [1]. Today, the advantages of multi-gate devices are investigated and studied [2, 3]. Among the various multi-gate devices, triple-gate (TG) devices appear more attractive, thanks to the superior ability of gate to control the conductive channel, compatibility to standard bulk planar and SOI CMOS processing, feasibility for digital/analog circuit applications, high current-driving capability, and low power consumption [4].

To utilize these advantages in MESFET devices, we proposed the triple-gate SOI MESFETs. To demonstrate the superior subthreshold performance of the proposed device, we present an analytical model which can describe the electrical parameters of the device including surface potential, threshold voltage, subthreshold swing, and short-channel effects. By using the analytical model, the specific features of the devices have been studied with satisfactory short-channel characteristics. The analytical model for potential distribution is derived by solving 3-D Poisson's equation using the superposition technique along with the separation of variables method. The obtained model is then used to find the expression of the threshold voltage and subthreshold swing. This model then offers a useful tool for characterization of the presented tri-gate MESFET. Also, the effect of device parameters is studied using the model obtained in this chapter. To validate the correctness of the analytical model, the results obtained by the analytical model are compared with the TCAD simulation results, and good accordance has been observed.

I. S. Amiri et al., *Device Physics, Modeling, Technology, and Analysis for Silicon MESFET*, https://doi.org/10.1007/978-3-030-04513-5_5

5.2 Device Structure

The structure of the proposed device is formed as the SOI formation in which a silicon channel laterally sits on top of a buried oxide layer, and Schottky gate electrode wraps around the silicon fin. The three-dimensional view of the proposed device structure used for analysis and simulation purpose is illustrated in Fig. 5.1.

As shown in the figure, the dimensions of the channel are indicated by L for channel length, H for channel height, and W for channel width. The Cartesian axes of x, y, and z are along the channel length, width, and height directions, respectively.

5.3 Definition of Channel Potential

The three-dimensional form of Poisson's equation of the Cartesian coordinate system can be written as:

$$\frac{\partial^2 \phi(x,y,z)}{\partial x^2} + \frac{\partial^2 \phi(x,y,z)}{\partial y^2} + \frac{\partial^2 \phi(x,y,z)}{\partial z^2} = -\frac{qN_D}{\varepsilon_{\text{si}}} \tag{5.1}$$

With $0 \leq x \leq L,\ -\frac{H}{2} \leq y \leq \frac{H}{2},\ -\frac{W}{2} \leq z \leq \frac{W}{2}$

where $\Phi(x,y,z)$ is the channel potential, q is electron charge, N_D is the donor concentration, and ε_{si} is the permittivity of silicon.

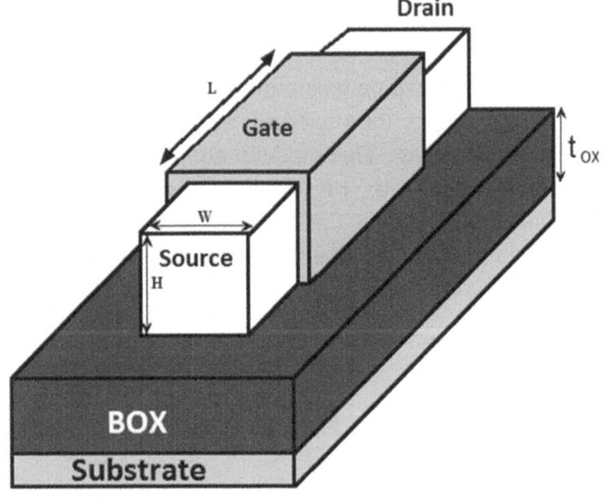

Fig. 5.1 The three-dimensional view of the proposed three-gate SOI MESFET

5.3.1 Boundary Conditions

Figure 5.2 shows the longitudinal and perpendicular cross-sections of an n-channel three-gate transistor, along with the axis and symbol conventions. From the figure, the boundary conditions for the device are described as follows:

(a) Top boundary condition

The boundary condition for top gate–silicon film interface can be expressed as:

$$\Phi(x, y, z)\big|_{y=-\frac{H}{2}} = V_{GS} - \Phi_{bi} \tag{5.2}$$

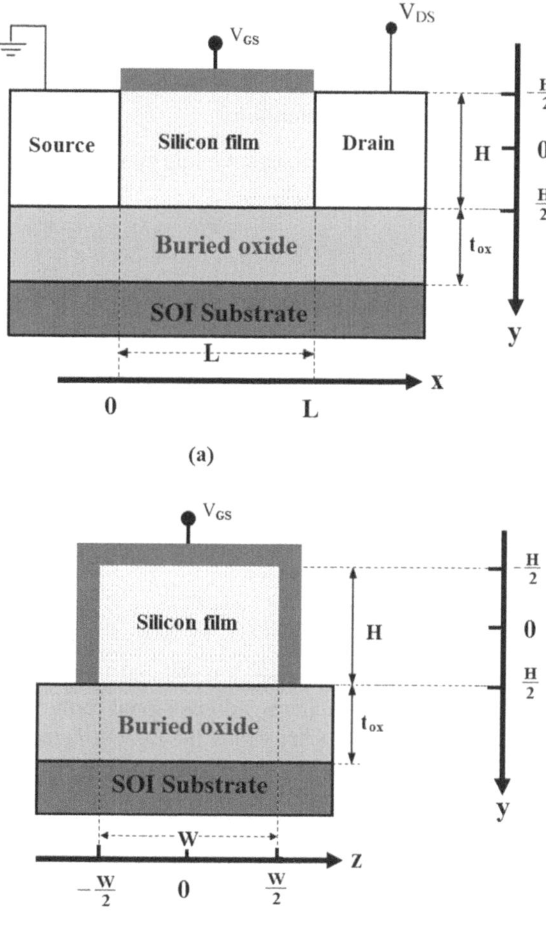

Fig. 5.2 Cross-section of the three-gate MESFET showing the boundary conditions and axes used in modeling: (**a**) longitudinal to current-flow direction, (**b**) and perpendicular to current-flow direction

(b) Left and Right boundary condition

As the side gates are connected to the top gate, for the interface of the side gates and silicon film we can write:

$$\Phi(x, y, z)|_{z=-\frac{W}{2}} = V_{GS} - \Phi_{bi} \tag{5.3}$$

$$\Phi(x, y, z)|_{z=\frac{W}{2}} = V_{GS} - \Phi_{bi} \tag{5.4}$$

(c) Bottom boundary condition

The bottom boundary belongs to the border between the silicon film and buried oxide layer. The electric field at this edge is continuous which is described as:

$$\left.\frac{d\Phi(x, y, z)}{dy}\right|_{y=\frac{H}{2}} + \frac{c_{ox}}{\varepsilon_{si}}\Phi(x, y, z)|_{y=\frac{H}{2}} = \frac{c_{ox}}{\varepsilon_{si}}(V_{GS} - V_{fb}) \tag{5.5}$$

(d) Source boundary condition

The boundary condition for source end can be expressed as:

$$\Phi(x, y, z)|_{x=0} = V_{bi} \tag{5.6}$$

(e) Drain boundary condition

Also for drain terminal, we can write

$$\Phi(x, y, z)|_{x=L} = V_{DS} + V_{bi} \tag{5.7}$$

To solve the Poisson's equation, we used the separation of variables method along with the superposition technique. According to this, we separated the potential $\Phi(x,y,z)$ into three parts as:

$$\Phi(x, y, z) = \Phi^{1D}(y) + \Phi^{2D}(x, y) + \Phi^{3D}(x, y, z) \tag{5.8}$$

Hence, $\Phi^{1D}(y)$ is the one-dimensional form of Poisson's equation that depends only on y. $\Phi^{2D}(x,y)$ is the two-dimensional form of Laplace's equation that depends on x and y. Finally, $\Phi^{3D}(x,y,z)$ is the three-dimensional form of Laplace's equation that depends on x, y, and z. Now, the resultant solution of (5.8) can be considered as the sum of 1D, 2D, and 3D potential solutions. Regarding this, we need to solve $\Phi^{1D}(y)$, $\Phi^{2D}(x,y)$, and $\Phi^{3D}(x,y,z)$ separately.

5.3.2 1-D Poisson's Equation for $\Phi^{1D}(y)$

One-dimensional Poisson's equation for $\Phi^{1D}(y)$ is expressed as:

$$\frac{\partial^2 \Phi^{1D}(y)}{\partial y^2} = -\frac{qN_D}{\varepsilon_{si}} - \frac{H}{2} \le y \le \frac{H}{2} \tag{5.9}$$

Regarding Eqs. (5.2) and (5.5), the boundary conditions for $\Phi^{1D}(y)$ are defined as:

$$\Phi^{1D}(y)\big|_{y=-\frac{H}{2}} = V_{GS} - \Phi_{bi} \tag{5.10}$$

$$\left.\frac{d\Phi^{1D}(y)}{dy}\right|_{y=\frac{H}{2}} + \frac{c_{ox}}{\varepsilon_{si}}\Phi^{1D}(y)\big|_{y=\frac{H}{2}} = \frac{c_{ox}}{\varepsilon_{si}}\left(V_{GS} - V_{fb}\right) \tag{5.11}$$

Since $\Phi^{1D}(y)$ depends on one variable (y), the partial derivatives can be replaced by ordinary derivatives, and the equation can be integrated directly. Integrating both sides of Eq. (5.9), we have

$$\frac{d\Phi^{1D}(y)}{dy} = -\frac{qN_D}{\varepsilon_{si}}y + c_1 \tag{5.12}$$

Integrating again, we get

$$\Phi^{1D}(y) = -\frac{qN_D}{2\varepsilon_{si}}y^2 + c_1 y + c_2 \tag{5.13}$$

To calculate the integration constants of c_1 and c_2, we apply the boundary conditions (5.10) and (5.11) into Eq. (5.13). This leads us to:

$$c_1 = \frac{2c_{ox}\left(\Phi_{bi} - V_{fb}\right) + qN_D H}{2d} \tag{5.14}$$

$$c_2 = V_{GS} + \frac{qN_D H^2(2\varepsilon_{si} + d) - 4\varepsilon_{si}\left[c_{ox}HV_{fb} + \Phi_{bi}(\varepsilon_{si} + d)\right]}{8\varepsilon_{si}d} \tag{5.15}$$

with:

$$d = c_{ox}H + \varepsilon_{si} \tag{5.16}$$

5.3.3 2-D Laplace's Equation for Φ^{2D}(x,y)

This section aims to solve the two-dimensional Laplace's equation in the Cartesian coordinate system in the following form:

$$\frac{\partial^2 \Phi^{2D}(x,y)}{\partial x^2} + \frac{\partial^2 \Phi^{2D}(x,y)}{\partial y^2} = 0 \quad 0 \le x \le L - \frac{H}{2} \le y \le \frac{H}{2} \tag{5.17}$$

The boundary conditions of $\Phi^{2D}(x, y)$ are defined as follows:

$$\Phi^{2D}(x,y)\big|_{y=-\frac{H}{2}} = 0 \tag{5.18}$$

$$\left.\frac{d\Phi^{2D}(x,y)}{dy}\right|_{y=\frac{H}{2}} + \frac{c_{ox}}{\varepsilon_{si}} \Phi^{2D}(x,y)\big|_{y=\frac{H}{2}} = 0 \tag{5.19}$$

$$\Phi^{2D}(x,y)\big|_{x=0} = V_{bi} - \Phi^{1D}(y) \tag{5.20}$$

$$\Phi^{2D}(x,y)\big|_{x=L} = V_{DS} + V_{bi} - \Phi^{1D}(y) \tag{5.21}$$

$\Phi^{1D}(y)$ in the above equations is represented by Eq. (5.13). Now, we use the method of separation of variables and suppose that $\Phi^{2D}(x,y)$ is expressible as the product of a function of x alone and a function of y alone [5]. Replacing the function of x by $X(x)$ and the function of y by $Y(y)$, $\Phi^{2D}(x,y)$ can be expressed as the product of two separate solutions [6] as follows:

$$\Phi^{2D}(x,y) = X(x) \cdot Y(y) \tag{5.22}$$

where $X(x)$ is only dependent on the x variable, and $Y(y)$ on the y variable. The property of dependence on a variable alone provides the separation of Laplace's equation into two scalar equations which are dependent on a distinct variable. Substitution of (5.22) into (5.17) gives

$$X(x)\frac{d^2Y(y)}{dy^2} + Y(y)\frac{d^2X(x)}{dx^2} = 0 \tag{5.23}$$

Considering $X(x) \times Y(y) \neq 0$ and dividing both sides by $X(x) \times Y(y)$, we obtain

$$-\frac{1}{Y(y)}\frac{d^2Y(y)}{dy^2} = \frac{1}{X(x)}\frac{d^2X(x)}{dx^2} = \lambda^2 \tag{5.24}$$

Note that in Eq. (5.24), the left side doesn't depend on x as the right side doesn't depend on y. We conclude that both sides of the equation cannot be a function of x and y. That means both of these quantities must be constant to have a logical separation. Now, Eq. (5.24) can be separated into two ordinary differential equations as follows:

$$\frac{d^2Y(y)}{dy^2} + \lambda^2 Y(y) = 0 \tag{5.25}$$

$$\frac{d^2X(x)}{dx^2} - \lambda^2 X(x) = 0 \tag{5.26}$$

Since the above equations are independent, we can solve them separately. Replacing Eq. (5.22) into the boundary conditions (5.18), (5.19), (5.20), and (5.21) gives us:

$$X(x)Y\left(\frac{-H}{2}\right) = 0 \tag{5.27}$$

$$X(x)\frac{dY\left(\frac{H}{2}\right)}{dy} + \frac{c_{ox}}{\varepsilon_{si}}X(x)Y\left(\frac{H}{2}\right) = 0 \tag{5.28}$$

$$X(0)Y(y) = V_{bi} - \Phi^{1D}(y) \tag{5.29}$$

$$X(L)Y(y) = V_{DS} + V_{bi} - \Phi^{1D}(y) \tag{5.30}$$

Assuming that λ^2 is positive, the general solution for $Y(y)$ is obtained as:

$$Y(y) = A\sin(\lambda y) + B\cos(\lambda y) \tag{5.31}$$

Here, we know that the solution of equations in the form of (5.25) and (5.26) can also be in the form of spatial sine, spatial cosine, hyperbolic sine, and hyperbolic cosine functions. Also depending on the problem, either the exponential or harmonic form can be used. To obtain a specific solution, the constants A and B must be attained by using the derived boundary conditions. By differentiating on (5.31) over y, we have

$$\frac{dY(y)}{dy} = \lambda A\cos(\lambda y) - \lambda B\sin(\lambda y) \tag{5.32}$$

By applying Eqs. (5.27) and (5.28) into (5.31) and (5.32), we have

$$-A\sin\left(\lambda\frac{H}{2}\right) + B\cos\left(\lambda\frac{H}{2}\right) = 0 \tag{5.33}$$

$$A\lambda\cos\left(\lambda\frac{H}{2}\right) - B\lambda\sin\left(\lambda\frac{H}{2}\right) + \frac{c_{ox}}{\varepsilon_{si}}\left[A\sin\left(\lambda\frac{H}{2}\right) + B\cos\left(\lambda\frac{H}{2}\right)\right] = 0 \tag{5.34}$$

Using Eqs. (5.33) and (5.34), we obtain

$$\tan\left(\lambda\frac{H}{2}\right) = \frac{2c_{ox}}{\lambda\varepsilon_{si}} \tag{5.35}$$

According to the fact that tangent function is periodic with period π, Eq. (5.35) has infinite solutions. Here, we use the subscript of "n" to indicate each solution. In this way, the values λ_n are the eigenvalues, and the function $Y(y)$ is the eigenfunction of Eq. (5.31). It is evident that for each value of λ_n, the values of A and B in $Y(y)$ will be different. Therefore, we rewrite Eq. (5.31) as:

$$Y_n(y) = A_n \sin(\lambda_n y) + B_n \cos(\lambda_n y) \tag{5.36}$$

By replacing B_n from Eq. (5.33), we have

$$Y_n(y) = A_n\left[\sin(\lambda_n y) + \frac{2c_{ox}}{\lambda\varepsilon_{si}}\cos(\lambda_n y)\right] \tag{5.37}$$

Refer to Eq. (5.26). The general solution for $X(x)$ is

$$X(x) = C' \cdot \exp(\lambda_n(x-L)) + D' \cdot \exp(-\lambda_n x) \tag{5.38}$$

where C' and D' are integration constants that must be found. According to (5.22), (5.37), and (5.38), the solution of $\Phi^{2D}(x,y)$ for all values of λ_n is

$$\begin{aligned}\Phi_n^{2D}(x,y,\lambda_n) &= \left[C_n'\exp(\lambda_n(x-L)) + D_n'\exp(-\lambda_n x)\right] \\ &\quad\times A_n\left[\sin\left(\lambda_n y + \frac{2c_{ox}}{\lambda\varepsilon_{si}}\right)\cos(\lambda_n y)\right]\end{aligned} \tag{5.39}$$

Note that Eq. (5.17) is linear and homogeneous. According to this, the sum of all solutions of this equation is also a solution to it. Therefore, we can write the complete solution of (5.17) as:

$$\Phi^{2D}(x,y) = \sum\nolimits_{n=1}^{\infty}\left[C_n\exp(\lambda_n(x-L)) + D_n\exp(-\lambda_n x)\right]\left[\sin(\lambda_n y) + \frac{2c_{ox}}{\lambda\varepsilon_{si}}\cos(\lambda_n y)\right] \tag{5.40}$$

where $C_n = A_n \times C_n'$ and $D_n = A_n \times D_n'$. In the next step, we must calculate the arbitrary constants of C_n and D_n. For this purpose, we apply boundary conditions (5.20) and (5.21). It leads us to:

$$\Phi^{2D}(0,y) = \sum_{n=1}^{\infty}\left[C_n \exp(-\lambda_n L) + D_n\right]\left[\sin(\lambda_n y) + \frac{2c_{ox}}{\lambda\varepsilon_{si}}\cos(\lambda_n y)\right]$$

$$= V_{bi} + \frac{qN_D}{\varepsilon_{si}}y^2 - c_1 y - c_2 \tag{5.41}$$

$$\Phi^{2D}(L,y) = \sum_{n=1}^{\infty}\left[C_n + D_n \exp(-\lambda_n L)\right]\left[\sin(\lambda_n y) + \frac{2c_{ox}}{\lambda\varepsilon_{si}}\cos(\lambda_n y)\right]$$

$$= V_{DS} + V_{bi} + \frac{qN_D}{\varepsilon_{si}}y^2 - c_1 y - c_2 \tag{5.42}$$

Now, we multiply both Eqs. (5.41) and (5.42) by $[\sin(\lambda_m y) + (2c_{ox}/\lambda\varepsilon_{si})\cos(\lambda_m y)]$ and use the property of orthogonality of the sine function. Now, integrating these equations concerning y, we have

$$\int_{-\frac{H}{2}}^{\frac{H}{2}} \Phi^{2D}(0,y)\left[\sin(\lambda_m y) + \frac{2c_{ox}}{\lambda\varepsilon_{si}}\cos(\lambda_m y)\right]dy$$

$$= \int_{-\frac{H}{2}}^{\frac{H}{2}} \sum_{n=1}^{\infty}\left[C_n \exp(-\lambda_n L) + D_n\right]$$

$$\times\left[\sin(\lambda_n y) + \frac{2c_{ox}}{\lambda\varepsilon_{si}}\cos(\lambda_n y)\right].\left[\sin(\lambda_m y) + \frac{2c_{ox}}{\lambda\varepsilon_{si}}\cos(\lambda_m y)\right]dy \tag{5.43}$$

$$\int_{-\frac{H}{2}}^{\frac{H}{2}} \Phi^{2D}(L,y)\left[\sin(\lambda_m y) + \frac{2c_{ox}}{\lambda\varepsilon_{si}}\cos(\lambda_m y)\right]dy$$

$$= \int_{-\frac{H}{2}}^{\frac{H}{2}} \sum_{n=1}^{\infty}\left[C_n + D_n \exp(-\lambda_n L)\right]$$

$$\times\left[\sin(\lambda_n y) + \frac{2c_{ox}}{\lambda\varepsilon_{si}}\cos(\lambda_n y)\right].\left[\sin(\lambda_m y) + \frac{2c_{ox}}{\lambda\varepsilon_{si}}\cos(\lambda_m y)\right]dy \tag{5.44}$$

As both integration and summation functions are independent, we can exchange them on the RHS:

$$\int_{-\frac{H}{2}}^{\frac{H}{2}} \Phi^{2D}(0,y)\left[\sin(\lambda_n y) + \frac{2c_{ox}}{\lambda\varepsilon_{si}}\cos(\lambda_n y)\right]dy$$

$$= \left[C_n \exp(-\lambda_n L) + D_n\right] \times \int_{-\frac{H}{2}}^{\frac{H}{2}}\left[\sin(\lambda_n y) + \frac{2c_{ox}}{\lambda\varepsilon_{si}}\cos(\lambda_n y)\right]^2 dy m = n \tag{5.45}$$

$$\int_{-\frac{H}{2}}^{\frac{H}{2}} \Phi^{2D}(L,y)\left[\sin(\lambda_n y)+\frac{2c_{ox}}{\lambda\varepsilon_{si}}\cos(\lambda_n y)\right]dy$$
$$= [C_n + D_n \exp(-\lambda_n L)] \times \int_{-\frac{H}{2}}^{\frac{H}{2}} \left[\sin(\lambda_n y)+\frac{2c_{ox}}{\lambda\varepsilon_{si}}\cos(\lambda_n y)\right]^2 dy m = n \quad (5.46)$$

or

$$C_n \exp(-\lambda_n L) + D_n = \frac{\beta_n}{\alpha_n} \quad (5.47)$$

$$C_n + D_n \exp(-\lambda_n L) = \frac{\gamma_n}{\alpha_n} \quad (5.48)$$

With:

$$\alpha_n = \int_{-\frac{H}{2}}^{\frac{H}{2}} \left[\sin(\lambda_n y)+\frac{2c_{ox}}{\lambda\varepsilon_{si}}\cos(\lambda_n y)\right]^2 dy \quad (5.49)$$

$$\beta_n = \int_{-\frac{H}{2}}^{\frac{H}{2}} \Phi^{2D}(0,y)\left[\sin(\lambda_n y)+\frac{2c_{ox}}{\lambda\varepsilon_{si}}\cos(\lambda_n y)\right]dy \quad (5.50)$$

$$\gamma_n = \int_{-\frac{H}{2}}^{\frac{H}{2}} \Phi^{2D}(L,y)\left[\sin(\lambda_n y)+\frac{2c_{ox}}{\lambda\varepsilon_{si}}\cos(\lambda_n y)\right]dy \quad (5.51)$$

Evaluating the integrals, the parameters α_n, β_n, and γ_n are calculated as:

$$\alpha_n = \frac{1}{2\lambda_n}\left[(p_n^2-1)\sin(\lambda_n H) + (p_n^2+1)\lambda_n H\right] \quad (5.52)$$

$$\beta_n = \frac{1}{2\lambda_n^3}\left\{r_n \sin\left(\lambda_n \frac{H}{2}\right) + 2\lambda_n H\left[c_1\lambda_n + \frac{qN_D}{\varepsilon_{si}}p_n\right]\cos\left(\lambda_n\frac{H}{2}\right)\right\} \quad (5.53)$$

$$\gamma_n = \frac{1}{2\lambda_n^3}\left\{\left[r_n + 4p_n\lambda_n^2 V_{DS}\right]\sin\left(\lambda_n \frac{H}{2}\right) + 2\lambda_n H\left[c_1\lambda_n + \frac{qN_D}{\varepsilon_{si}}p_n\right]\cos\left(\lambda_n\frac{H}{2}\right)\right\} \quad (5.54)$$

with:

$$p_n = \frac{2c_{ox}}{\lambda_n \varepsilon_{si}} \quad (5.55)$$

$$r_n = p_n\left(\frac{qN_D}{2\varepsilon_{si}}H^2 - 4c_2 + 4v_{bi}\right)\lambda_n^2 - 4c_1\lambda_n - 4p_n\frac{qN_D}{\varepsilon_{si}} \tag{5.56}$$

Now, C_n and D_n are defined as:

$$C_n = \frac{-\beta_n + \gamma_n e^{\lambda_n L}}{2\alpha_n \sin h(\lambda_n L)} \tag{5.57}$$

$$D_n = \frac{-\gamma_n + \beta e^{\lambda_n L}}{2\alpha_n \sin h(\lambda_n L)} \tag{5.58}$$

5.3.4 3-D Laplace's Equation for Φ^{3D}(x,y,z)

In this section, we involved with the solution of Laplace's equation which varies with three coordinates as below:

$$\frac{\partial^2\Phi^{3D}(x,y,z)}{\partial x^2} + \frac{\partial^2\Phi^{3D}(x,y,z)}{\partial y^2} + \frac{\partial^2\Phi^{3D}(x,y,z)}{\partial z^2} = 0$$

$$0 \le x \le L - \frac{H}{2} \le y \le \frac{H}{2} - \frac{W}{2} \le z \le \frac{W}{2} \tag{5.59}$$

Regarding (5.2)–(5.8), we defined the following boundary conditions for $\Phi^{3D}(x, y,z)$:

$$\Phi^{3D}(x,y,z)\Big|_{y=-\frac{H}{2}} = 0 \tag{5.60}$$

$$\left.\frac{d\Phi^{3D}(x,y,z)}{dy}\right|_{y=\frac{H}{2}} + \frac{c_{ox}}{\varepsilon_{si}}\Phi^{3D}(x,y,z)\Big|_{y=\frac{H}{2}} = 0 \tag{5.61}$$

$$\Phi^{3D}(x,y,z)\Big|_{z=-\frac{W}{2}} = V_{GS} - \Phi_{bi} - \Phi^{1D}(y) - \Phi^{2D}(x,y) \tag{5.62}$$

$$\Phi^{3D}(x,y,z)\Big|_{z=\frac{W}{2}} = V_{GS} - \Phi_{bi} - \Phi^{1D}(y) - \Phi^{2D}(x,y) \tag{5.63}$$

$$\Phi^{3D}(x,y,z)\Big|_{x=0} = 0 \tag{5.64}$$

$$\Phi^{3D}(x,y,z)\Big|_{x=L} = 0 \tag{5.65}$$

To obtain the solution of Eq. (5.56), similar to the last section, we separated the variables again. According to this, the solution of $\Phi^{3D}(x,y,z)$ can be expressed as:

$$\Phi^{3D}(x, y, z) = X(x) \cdot Y(y) \cdot Z(z) \tag{5.66}$$

where $X(x)$ depends on only x, $Y(y)$ depends on only y, and $Z(z)$ depends on only z. Substituting (5.66) into Eq. (5.59) and dividing the result through by (5.66), we get

$$\frac{1}{X(x)}\frac{d^2X(x)}{dx^2} + \frac{1}{Y(y)}\frac{d^2Y(y)}{dy^2} + \frac{1}{Z(z)}\frac{d^2Z(z)}{dz^2} = 0 \tag{5.67}$$

or

$$-\left[\frac{1}{X(x)}\frac{d^2X(x)}{dx^2} + \frac{1}{Y(y)}\frac{d^2Y(y)}{dy^2}\right] = \frac{1}{Z(z)}\frac{d^2Z(z)}{dz^2} = k^2 \tag{5.68}$$

Accordingly, we can write

$$-\frac{1}{Y(y)}\frac{d^2Y(y)}{dy^2} = k^2 + \frac{1}{X(x)}\frac{d^2X(x)}{dx^2} = \lambda^2 \tag{5.69}$$

and

$$\frac{1}{X(x)}\frac{d^2X(x)}{dx^2} = -(k^2 - \lambda^2) = -h^2 \tag{5.70}$$

In this case, each term of (5.67) depends on a single variable. Thus, each term must be constant to have a logical separation. Therefore, we can rewrite the partial differential equation (PDE) regarding ordinary differential equation as:

$$\frac{1}{X(x)}\frac{d^2X(x)}{dx^2} = -h^2 \tag{5.71}$$

$$\frac{1}{Y(y)}\frac{d^2Y(y)}{dy^2} = -\lambda^2 \tag{5.72}$$

$$\frac{1}{Z(z)}\frac{d^2Z(z)}{dz^2} = k^2 \tag{5.73}$$

Note that all three equations are completely independent so that we can solve them separately. Replacing (5.66) into boundary conditions (5.60)–(5.65), we get

$$X(x)Y\left(\frac{-H}{2}\right)Z(z) = 0 \tag{5.74}$$

$$X(x)\frac{dY\left(\frac{H}{2}\right)}{dy}Z(z)+\frac{c_{ox}}{\varepsilon_{si}}X(x)Y\left(\frac{H}{2}\right)Z(z)=0 \tag{5.75}$$

$$X(0)Y(y)Z(z)=0 \tag{5.76}$$

$$X(L)Y(y)Z(z)=0 \tag{5.77}$$

$$X(x)Y(y)Z\left(\frac{-W}{2}\right)=V_{GS}-\Phi_{bi}-\Phi^{1D}(y)-\Phi^{2D}(x,y) \tag{5.78}$$

$$X(x)Y(y)Z\left(\frac{W}{2}\right)=V_{GS}-\Phi_{bi}-\Phi^{1D}(y)-\Phi^{2D}(x,y) \tag{5.79}$$

Hence, $X(x)$ and $Y(y)$ are the solutions of Eqs. (5.71) and (5.72), respectively. It is to be noted that Eqs. (5.78) and (5.79) don't be satisfied by some arbitrary values of $\Phi^{1D}(y)$ and $\Phi^{2D}(x,y)$. Therefore, we inevitably used Eqs. (5.62) and (5.63). To avoid complicated calculation, we used the simplified form of Eqs. (5.74) to (5.77) as follows:

$$Y\left(\frac{-H}{2}\right)=0 \tag{5.80}$$

$$\frac{dY\left(\frac{H}{2}\right)}{dy}+\frac{c_{ox}}{\varepsilon_{si}}Y\left(\frac{H}{2}\right)=0 \tag{5.81}$$

$$X(0)=0 \tag{5.82}$$

$$X(L)=0 \tag{5.83}$$

For $Y(y)$, we follow the similar solution to Eq. (5.36). It yields

$$Y_n(y)=A_n\left[\sin(\lambda_n y)+\frac{2c_{ox}}{\lambda\varepsilon_{si}}\cos(\lambda_n y)\right] \tag{5.84}$$

The parameter λ_n is given by:

$$\tan\left(\lambda_n\frac{H}{2}\right)=\frac{2c_{ox}}{\lambda_n\varepsilon_{si}} \tag{5.85}$$

A general solution for Eq. (5.68) is

$$X(x)=E\sin(hx)+F\cos(hx) \tag{5.86}$$

where E and F are integration constants. Applying the boundary conditions (5.82) and (5.83) yields

$$F = 0 \tag{5.87}$$

$$E \sin(hL) = 0 \tag{5.88}$$

Supposing $E \neq 0$, we obtain

$$h_m = \frac{m\pi}{L} \quad m = 1, 2, 3, \ldots \tag{5.89}$$

The solution of $X(x)$ can be written as:

$$X_m(x) = E_m \sin\left(\frac{m\pi}{L}\right) x \tag{5.90}$$

It is evident that $m > 0$ because $m = 0$ leads us to $M(x) = 0$. Also, $m < 0$ does not give us any extra information as E_m is arbitrarily constant. A general solution to $Z(z)$ is

$$Z(z) = M \cdot \exp\left(k\left(z - \frac{W}{2}\right)\right) + N \cdot \exp\left(-k\left(z + \frac{W}{2}\right)\right) \tag{5.91}$$

where M and N are integration constants that must be found. From Eq. (5.67), we have

$$k_{nm}^2 = \lambda_n^2 + h_m^2 = \lambda_n^2 + \left(\frac{m\pi}{L}\right)^2 \tag{5.92}$$

Thus, we can rewrite $Z(z)$ as:

$$Z_{nm}(z) = M_{nm} \cdot \exp\left(k_{nm}\left(z - \frac{W}{2}\right)\right) + N_{nm} \cdot \exp\left(-k_{nm}\left(z + \frac{W}{2}\right)\right) \tag{5.93}$$

The solution of $\Phi^{3D}(x,y,z)$ is described as:

$$\begin{aligned} \Phi^{3D}(x, y, z; m, n) = \; & E_m \sin\left(\frac{m\pi}{L} \cdot x\right) \times A_n \left[\sin(\lambda_n y) + \frac{2c_{ox}}{\lambda \varepsilon_{si}} \cos(\lambda_n y)\right] \\ & \times \left[M_{nm} \cdot \exp\left(k_{nm}\left(z - \frac{W}{2}\right)\right) + N_{nm} \cdot \exp\left(-k_{nm}\left(z + \frac{W}{2}\right)\right)\right] \end{aligned} \tag{5.94}$$

Since there are infinite solutions for the problem, a sum of all solutions also can be a solution. Furthermore, the unknown coefficients An and E_m are multiplied by M_{nm} and N_{nm} so that we can remove them from the equation. Therefore, Eq. (5.94) becomes

$$\Phi^{3D}(x,y,z) = \sum_{n=1}^{\infty}\sum_{m=1}^{\infty} \sin\left(\frac{m\pi}{L}\cdot x\right)\left[\sin(\lambda_n y) + \frac{2c_{ox}}{\lambda\varepsilon_{si}}\cos(\lambda_n y)\right] \times \left[M_{nm}\cdot \exp\left(k_{nm}\left(z-\frac{W}{2}\right)\right) + N_{nm}\cdot \exp\left(-k_{nm}\left(z+\frac{W}{2}\right)\right)\right] \tag{5.95}$$

Applying the boundary conditions (5.75) and (5.76), we have

$$\sum_{n=1}^{\infty}\sum_{m=1}^{\infty} \sin\left(\frac{m\pi}{L}\cdot x\right)\left[\sin(\lambda_n y) + \frac{2c_{ox}}{\lambda\varepsilon_{si}}\cos(\lambda_n y)\right] \times \left[M_{nm}\cdot \exp(-\mu_{nm}W) + N_{nm}\right] = V_{GS} - \Phi_{bi} - \Phi^{1D}(y) - \Phi^{2D}(x,y) \tag{5.96}$$

$$\sum_{n=1}^{\infty}\sum_{m=1}^{\infty} \sin\left(\frac{m\pi}{L}\cdot x\right)\left[\sin(\lambda_n y) + \frac{2c_{ox}}{\lambda\varepsilon_{si}}\cos(\lambda_n y)\right] \times \left[M_{nm} + N_{nm}\cdot \exp(-\mu_{nm}W)\right] = V_{GS} - \Phi_{bi} - \Phi^{1D}(y) - \Phi^{2D}(x,y) \tag{5.97}$$

It is clear from Eqs. (5.96) and (5.97) that $M_{nm} = N_{nm}$. Based on the orthogonality of Eq. (5.59), we multiply both sides of the above equations by $\sin\left(\frac{m\pi}{L}.x\right)$ and $\left[\sin(\lambda_n y) + \tan\left(\lambda_n\frac{H}{2}\right)\cos(\lambda_n y)\right]$. Then, we integrate them from 0 to L over x and integrate them again from $H/2$ to $H/2$ over y. It yields

$$\int_0^L \int_{-\frac{H}{2}}^{\frac{H}{2}} \left[\Phi^{3D}\left(x,y,\frac{W}{2}\right)\right] \cdot \sin\left(\frac{m\pi}{L}\cdot x\right)\left[\sin(\lambda_n y) + \frac{2c_{ox}}{\lambda\varepsilon_{si}}\cos(\lambda_n y)\right] dydx = \int_0^L \int_{-\frac{H}{2}}^{\frac{H}{2}} \sin^2\left(\frac{m\pi}{L}\cdot x\right)\left[\sin(\lambda_n y) + \frac{2c_{ox}}{\lambda\varepsilon_{si}}\cos(\lambda_n y)\right]^2 \times M_{nm}\left[1+\exp(-k_{nm}W)\right]dydx \tag{5.98}$$

$\Phi^{1D}(y)$ and $\Phi^{2D}(x,y)$ are given by Eqs. (5.13) and (5.40). Evaluating the integrals and solving (5.95) for M_{nm}, we have

$$M_{nm} = N_{nm} = \frac{u_{nm}}{v_n\left(1 + e^{-\mu_{nm}W}\right)} \tag{5.99}$$

u_{nm} and v_n are expressed as:

$$u_{nm} = \int_0^L \int_{-\frac{H}{2}}^{\frac{H}{2}} \left[\Phi^{3D}\left(x, y, \frac{W}{2}\right)\right] \cdot \sin\left(\frac{m\pi}{L}.x\right)\left[\sin(\lambda_n y) + \frac{2c_{ox}}{\lambda\varepsilon_{si}}\cos(\lambda_n y)\right] dydx \tag{5.100}$$

$$v_{nm} = \int_0^L \int_{-\frac{H}{2}}^{\frac{H}{2}} \left[\sin(\lambda_n y) + \frac{2c_{ox}}{\lambda\varepsilon_{si}}\cos(\lambda_n y)\right]^2 \sin^2\left(\frac{m\pi}{L}\cdot x\right) dydx \tag{5.101}$$

Replacing $\Phi^{3D}\left(x, y, \frac{W}{2}\right)$ from equation (5.63) into (5.100), we obtain

$$\begin{aligned} u_{nm} &= \int_0^L \int_{-\frac{H}{2}}^{\frac{H}{2}} \left[V_{GS} - \Phi_{bi} - \Phi^{1D}(y) - \Phi^{2D}(x, y)\right] \\ &\quad \cdot \sin\left(\frac{m\pi}{L}.x\right)\left[\sin(\lambda_n y) + \frac{2c_{ox}}{\lambda\varepsilon_{si}}\cos(\lambda_n y)\right] dydx \\ &= u_{nm1} - u_{nm2} \end{aligned} \tag{5.102}$$

where:

$$u_{nm1} = \int_0^L \int_{-\frac{H}{2}}^{\frac{H}{2}} \left[V_{GS} - \Phi_{bi} - \Phi^{1D}(y)\right] \cdot \sin\left(\frac{m\pi}{L}.x\right)\left[\sin(\lambda_n y) + \frac{2c_{ox}}{\lambda\varepsilon_{si}}\cos(\lambda_n y)\right] dydx \tag{5.103}$$

$$u_{nm2} = \int_0^L \int_{-\frac{H}{2}}^{\frac{H}{2}} \Phi^{2D}(x, y) \cdot \sin\left(\frac{m\pi}{L}.x\right)\left[\sin(\lambda_n y) + \frac{2c_{ox}}{\lambda\varepsilon_{si}}\cos(\lambda_n y)\right] dydx \tag{5.104}$$

Substituting $\Phi^{1D}(y)$ and $\Phi^{2D}(x, y)$ in the above equations and solving the integrals, we have

$$u_{nm1} = \frac{[1 - cos\,(m\pi)]L}{2m\pi\lambda_n^3} \left\{p_n\lambda_n^2\left[H^2 a + 4(V_{GS} - V_{bi} - c_2) - 4c_1\lambda_n - 8pa\right] \sin\left(\lambda_n\frac{H}{2}\right) + 2H(2pa + c_1\lambda_n)\cos\left(\lambda_n\frac{H}{2}\right)\right\} \tag{5.105}$$

$$u_{nm2} = \frac{m\pi L}{2\lambda_n e^{\lambda_n L}\left(L^2\lambda_n^2 + m^2\pi^2\right)}\left\{2\lambda_n\alpha_n e^{\lambda_n L}[D_n - C_n\cos(m\pi)] + (C_n - D_n\cos(m\pi))\right\} \tag{5.106}$$

$$v_{nm} = \frac{1}{2\lambda_n}\left[\left(p_n^2 - 1\right)\sin(\lambda_n H) + \left(p_n^2 + 1\right)\lambda_n H\right] \tag{5.107}$$

Eventually, the complete solution of three-dimensional potential distribution can be written as:

$$\begin{aligned}\Phi(x, y, z) &= \Phi^{1D}(y) + \Phi^{2D}(x, y) + \Phi^{3D}(x, y, z) \\ &= -\frac{qN_D}{2\varepsilon_{si}}y^2 + c_1 y + c_2 \\ &\quad + \sum_{n=1}^{\infty}[C_n\exp(\lambda_n(x - L)) + D_n\exp(-\lambda_n x)]\left[\sin(\lambda_n y) + \frac{2c_{ox}}{\lambda\varepsilon_{si}}\cos(\lambda_n y)\right] \\ &\quad + \sum_{n=1}^{\infty}\sum_{m=1}^{\infty}\sin\left(\frac{m\pi}{L}.x\right)\left[\sin(\lambda_n y) + \frac{2c_{ox}}{\lambda\varepsilon_{si}}\cos(\lambda_n y)\right] \\ &\quad \times \left[M_{nm}\cdot\exp\left(h_{nm}\left(z - \frac{W}{2}\right)\right) + N_{nm}.\exp\left(-h_{nm}\left(z + \frac{W}{2}\right)\right)\right]\end{aligned} \tag{5.108}$$

5.4 Definition of the Threshold Voltage

Similar to the definition used in Chap. 3, the threshold voltage of a small geometry three-gate SOI MESFET is defined as the voltage applied to the gate to make the whole channel fully depleted. From this point of view, the threshold voltage is governed by the bottom potential in minimum position. Also, the minimum barrier height is attained in the center of the channel (at $y = H/2$). Therefore, at threshold condition, the surface potential at ($x_{\min}$, $H/2$, 0) must be equal to zero.

$$V_{th} = V_{GS} \text{ when } \Phi\left(x_{\min}, -\frac{H}{2}, 0\right) = 0 \tag{5.109}$$

In the above equation, $x_{\min}$ is the lateral position of the minimum bottom potential, also is called virtual source, and is found by differentiating (5.108) concerning x at $y = H/2$ and $z = 0$ and solving the below equation:

$$\frac{d\Phi\left(x_{\min}, \frac{H}{2}, 0\right)}{dx} = 0 \tag{5.110}$$

It is difficult to solve (5.110) analytically, and accordingly, we used the bisection method to find $x_{\min}$. Replacing (5.109) in (5.5) concerning (5.8), (5.13), (5.40), and (5.92), we have

$$V_{th} = V_{th}^{0} - \Delta V_{th}^{L} - \Delta V_{th}^{W} \tag{5.111}$$

In the above equation, V_{th}^{0} which depends on both L and W is the threshold voltage of long and wide three-gate MESFET. On the other side, ΔV_{th}^{L} indicates the reduction of the threshold voltage due to short-channel effects. Here, ΔV_{th}^{L} varies with L but is independent of W. Finally, ΔV_{th}^{W} is the reduction of the threshold voltage by 3-D effects and varies with W [7]. Regarding (5.11), (5.19), and (5.58), we can write

$$V_{th}^{0} = V_{fb} + \frac{\varepsilon_{si}}{c_{ox}} \left.\frac{d\Phi^{1D}(y)}{dy}\right|_{y=\frac{H}{2}} = V_{fb} + \frac{\varepsilon_{si}}{c_{ox}} - \frac{qN_D}{2\varepsilon_{si}} H + c_1 \tag{5.112}$$

$$\Delta V_{th}^{L} = \frac{\varepsilon_{si}}{c_{ox}} \left.\frac{d\Phi^{2D}(x,y)}{dy}\right|_{\substack{x = x_{\min} \\ y = \frac{H}{2}}} = \sum_{n=1}^{\infty} \lambda_n \left[C_n \exp(\lambda_n (x_{\min} - L)) + D_n \exp(-\lambda_n x_{\min})\right] \times \left[\cos\left(\lambda_n \frac{H}{2}\right) - \frac{2c_{ox}}{\lambda_n \varepsilon_{si}} \sin\left(\lambda_n \frac{H}{2}\right)\right] \tag{5.113}$$

$$\Delta V_{th}^{W} = \frac{\varepsilon_{si}}{c_{ox}} \left.\frac{d\Phi^{3D}(x,y,z)}{dy}\right|_{\substack{x = x_{\min} \\ y = \frac{H}{2} \\ z = 0}} = 2 \sum_{n=1}^{\infty} \sum_{m=1}^{\infty} \lambda_n M_{nm} \exp\left(-\mu_{nm} \frac{W}{2}\right) \sin\left(\frac{m\pi}{L} x_{\min}\right) \times \left[\cos\left(\lambda_n \frac{H}{2}\right) - \frac{2c_{ox}}{\lambda_n \varepsilon_{si}} \sin\left(\lambda_n \frac{H}{2}\right)\right] \tag{5.114}$$

5.5 Definition of the Subthreshold Swing

In the subthreshold condition, the concentration of carriers is low, and the drift current in the channel is negligible. Consequently, we can consider only diffusion current for modeling of subthreshold current [8]. The diffusion current density can be approximately described as [9]:

$$J_n(y,z) \cong qD_n \frac{n(x_{\min},y,z)}{L_e(y,z)}\left(1-e^{-\frac{V_{DS}}{V_t}}\right) \tag{5.115}$$

D_n is diffusion coefficient of the electrons, V_t is the thermal voltage, and $n(x_{\min},y,z)$ is the electron density at the virtual source and given by:

$$n(x_{\min},y,z) = \frac{n_i^2}{N_D}\exp\left(\frac{\Phi(x_{\min},y,z)}{V_t}\right) \tag{5.116}$$

The effective length of the channel is calculated as $L_e \equiv L - L_s - L_d + 2L_D$, where L_s and L_d represent the extension of the depletion regions of the source and drain junctions. They are estimated using the solution for channel potential as follows [10]:

$$L_s = \frac{2[\Phi_{bi} - \Phi(\min)]}{\left|\frac{\partial(x,y_{\min},z)}{\partial x}\right|_{\substack{x=0\\ z=0}}} \tag{5.117}$$

$$L_d = \frac{2[\Phi_{bi} + V_{DS} - \Phi(\min)]}{\left|\frac{\partial(x,y_{\min},z)}{\partial x}\right|_{\substack{x=L\\ z=0}}} \tag{5.118}$$

Φ (min) is the minimum surface potential at $x = x_{\min}$ and $z = 0$. Also, L_D is called Debye length and is calculated as:

$$L_D = \sqrt{\frac{\varepsilon_{si}V_t}{qN_D}} \tag{5.119}$$

Also, the minimum potential point along the y direction, $y_{\min}$, can be calculated by solving the below equation:

$$\left.\frac{\partial\Phi(x_{\min},y,0)}{dy}\right|_{y=y_{\min}} = 0 \tag{5.120}$$

Also, for evaluating the subthreshold current, we can write

$$I_{DS} = \int_{-\frac{W}{2}}^{\frac{W}{2}} \int_{-\frac{H}{2}}^{\frac{H}{2}} J_n(y,z)dydz$$
$$= \frac{qD_n n_i^2}{N_D L_e}[1 - \exp(-V_{DS}/V_t)] \int_{-\frac{W}{2}}^{\frac{W}{2}} \int_{-\frac{H}{2}}^{\frac{H}{2}} \exp\left[\frac{\Phi(x_{\min}, y, z)}{V_t}\right] dydz \quad (5.121)$$

Finally, the subthreshold swing is found as:

$$SS = \left\{\frac{\partial log(I_D)}{\partial V_{GS}}\right\}^{-1} \quad (5.122)$$

References

1. H. Mohammadi, H. Abdullah, C.F. Dee, A review on modeling the channel potential in multi-gate MOSFETs. Sains Malays. **43**, 861–866 (2014)
2. I. Ferain, C.A. Colinge, J.-P. Colinge, Multigate transistors as the future of classical metal-oxide-semiconductor field-effect transistors. Nature **479**, 310–316 (2011)
3. J.P. Colinge, Multigate Transistors: Pushing Moore's Law to the Limit, in *2014 International Conference on Simulation of Semiconductor Processes and Devices (SISPAD)*, 2014, pp. 313–316
4. P. Vimala, N.B. Balamurugan, New analytical model for nanoscale tri-gate SOI MOSFETs including quantum effects. IEEE J. Electron Device Soc. **2**, 1–7 (2014)
5. N. Ida, *Engineering Electromagnetics* (Springer, Cham, 2000)
6. P. Rakesh Kumar, S. Mahapatra, Analytical modeling of quantum threshold voltage for triple gate MOSFET. Solid State Electron. **54**, 1586–1591 (2010)
7. G. Katti, N. DasGupta, A. DasGupta, Threshold voltage model for mesa-isolated small geometry fully depleted SOI MOSFETs based on analytical solution of 3-D Poisson's equation. IEEE Trans. Electron Devices **51**, 1169–1177 (2004)
8. S. Tripathi, V. Narendar, A three-dimensional (3D) analytical model for subthreshold characteristics of uniformly doped FinFET. Superlattice. Microst. **83**, 476–487 (2015)
9. D.S. Havaldar, G. Katti, N. DasGupta, A. DasGupta, Subthreshold current model of FinFETs based on analytical solution of 3-D Poisson's equation. IEEE Trans. Electron Devices **53**, 737–742 (2006)
10. Y. Ping Chin, J.G. Fossum, Physical subthreshold MOSFET modeling applied to viable design of deep-submicrometer fully depleted SOI low-voltage CMOS technology. IEEE Trans. Electron Devices **42**, 1605–1613 (1995)

Chapter 6
Analytical Investigation of Subthreshold Performance of SOI MESFET Devices

6.1 Introduction

In the present chapter, the subthreshold performance of SOI MESFET devices, is analytically investigated. Moreover, the advantages of the new SOI MESFET designs over their classical counterpart are studied, and the obtained results are reported. For verifying the accuracy of the recommended analytical models in this work, all models are examined with the TCAD simulator ATLAS from SILVACO. The devices under study are simulated using conventional drift-diffusion model for the carrier transport along with the classical Fermi–Dirac statistics for the carrier distribution. We apperceive that model results may deviate from simulation results a little, mainly if the channel length is short. The discrepancies may originate from the full depletion estimate, and other approximations have been used.

6.2 Analysis of Short-Channel SOI MESFET

Regarding the expressions obtained in Chap. 3, the profile of bottom potential, threshold voltage, and the subthreshold swing of the device are calculated and plotted. In our calculation, 50 terms of Fourier series are used. Also, the TCAD ATLAS device simulator is used to examine the correctness of the analytical results. In all figures, solid lines correspond to the analytical results, while symbols indicate the simulation results.

I. S. Amiri et al., *Device Physics, Modeling, Technology, and Analysis for Silicon MESFET*, https://doi.org/10.1007/978-3-030-04513-5_6

6.2.1 Device Dimension and Parameters

Unless specified otherwise, the device parameters, some constants, and terminal voltages used in the analysis and simulation are listed in Table 6.1.

6.2.2 Analysis of the Channel Potential

In this section, the profile of bottom potential versus normalized position along the channel at different drain biases and channel lengths produced by our model, the ATLAS, and the model presented by Chiang are plotted. Figure 6.1 illustrates the plot for the channel length of $L = 100$ nm and three values of drain voltage ($V_{DS} = 0$, $V_{DS} = 0.5$, and $V_{DS} = 1$ V). The figure points out that the minimum bottom potential rises with increasing of drain voltage, whereas the position of minimum bottom potential remains nearly constant. In comparison with the results obtained by Chiang, the results computed from our model are more consistent with the simulation results. In Fig. 6.2, the plot for a fixed drain voltage ($V_{DS} = 0$ V) and two values of the gate length ($L = 100$ and $L = 500$ nm) are shown. It is evident from the figure that for a long-channel device ($L = 500$ nm), the channel potential is almost flat. The fields from source and drain cause the region of minimum potential has slenderized from a wider region. In a long-channel ($L = 500$ nm), into a minimum point near the center of the channel in short channel ($L = 100$ nm). Also, decreasing the gate length elevates the minimum bottom potential, due to the SCEs. Further, the position of minimum channel potential also shifts toward the source due to the drain-induced barrier lowering (DIBL). The excellent agreement of the minimum potential found by our model with ATLAS shows that the DIBL effect predicted by our model is far more exact than that by the last model presented by Chiang.

Table 6.1 Parameters used for analysis and simulation of short-channel SOI MESFET

Specified parameters	Symbol	Value
Channel doping concentration	N_D	5×10^{22} cm^{-3}
Source/Drain region's doping concentration	$N_{S/D}$	10^{20} cm^{-3}
Thickness of buried oxide layer	t_{ox}	400 nm
Thickness of silicon film	t_{si}	40 nm
Permittivity of silicon	ε_{si}	$11.8 \times 8.85 \times 10^{-14}$ F/cm
Permittivity of SiO_2	ε_{ox}	$3.9 \times 8.85 \times 10^{-14}$ F/cm
Built-in voltage of Schottky junction	Φ_{bi}	0.7 V
Gate length	L	100 nm
The flat-band voltage of SiO_2	V_{fb}	−0.r
Source/Drain doping concentration	$N_{S/D}$	10^{20} cm^{-3}
Midgap metal gate work-function	WF	4.71 eV

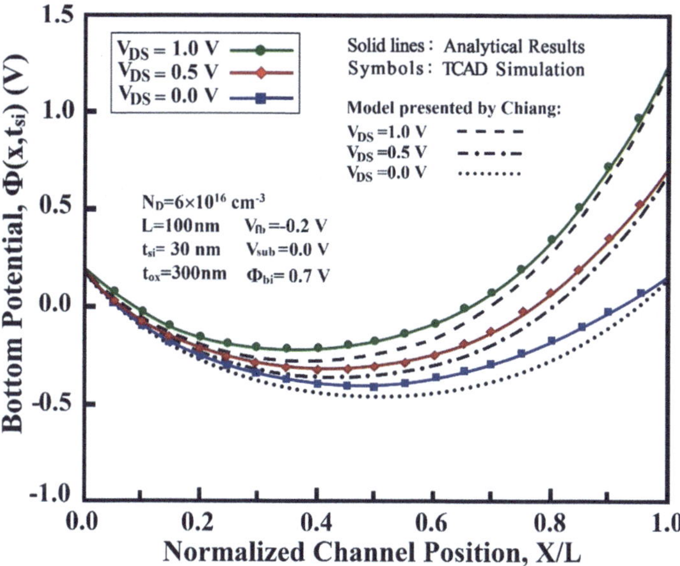

Fig. 6.1 The profile of bottom potential versus normalized channel position for $L = 100$ at three values of drain voltage obtained by our model (solid lines), the model presented by Chiang (dashed lines), and ATLAS simulation results (symbols)

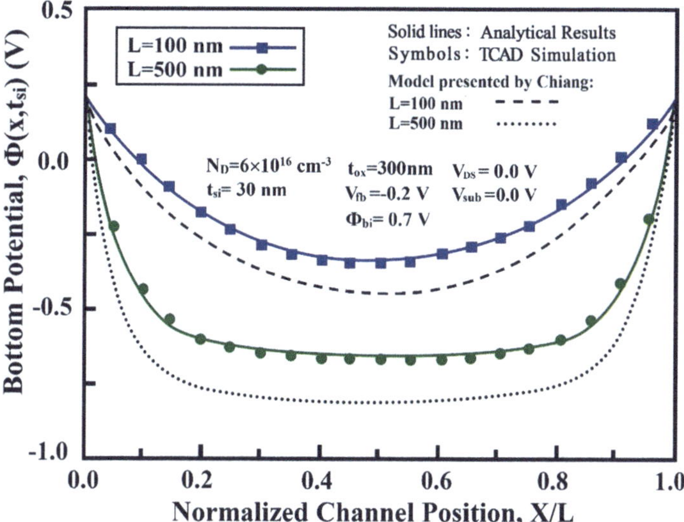

Fig. 6.2 The profile of bottom potential versus normalized channel position for $L = 100$ and 500 nm at $V_{DS} = 0$ obtained by our model (solid lines), the model presented by Chiang (dashed lines), and ATLAS simulation results (symbols)

6.2.3 Analysis of Threshold Voltage

The next interest is to plot the variation of threshold voltage against channel length attained analytically and its evaluation with ATLAS simulated results for different values of silicon film thickness, channel doping concentration, drain voltage, and substrate voltage. Figure 6.3 shows the profile of threshold voltage for three values of silicon film thickness ($t_{si} = 40$, $t_{si} = 80$, and $t_{si} = 100$ nm) in which the oxide thickness and channel doping concentration are fixed. It is evident from the figure that the thinner silicon film declines the threshold voltage shift and consequently suppresses the DIBL effect. The ultra-thin silicon film is one of the solutions used to protect the performance of the small geometry device from the SCEs as the device dimensions are continuously scaled down. For the device with short-channel length and the thick silicon film, the threshold voltage changes into a negative value, and the MESFET operates as a normally-on device. In comparison with the models presented by Chiang and Jit, our model is more consistent with the ATLAS simulation results.

Channel doping density is another effective device parameter for governing the threshold voltage. Figure 6.4 sketches the threshold voltages versus the channel length for three values of channel doping concentration ($N_D = 1 \times 10^{16}$, $N_D = 3 \times 10^{16}$, and $N_D = 5 \times 10^{16}$ cm^{-3}) as a device parameter in which $t_s = 100$ nm

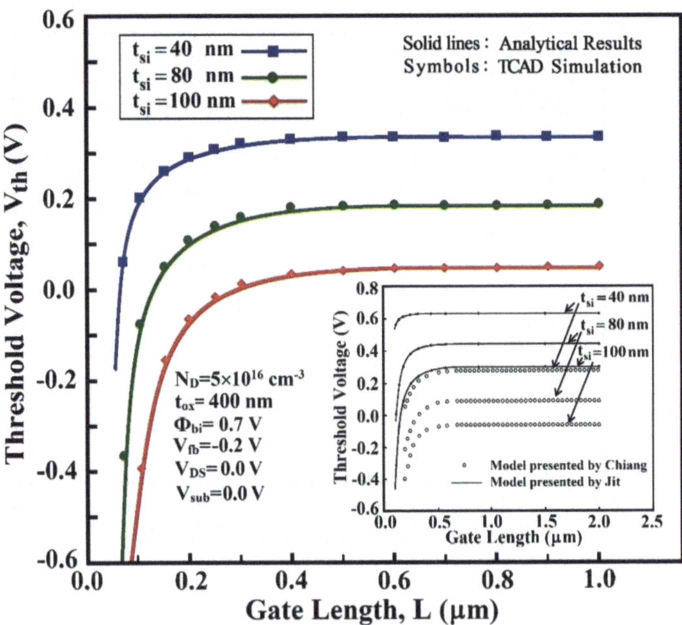

Fig. 6.3 The plot of threshold voltage versus the channel length for various silicon film thicknesses. Solid lines correspond to the model, and symbols correspond to the simulation results. (Inset) The analytical results obtained by Chiang and Jit

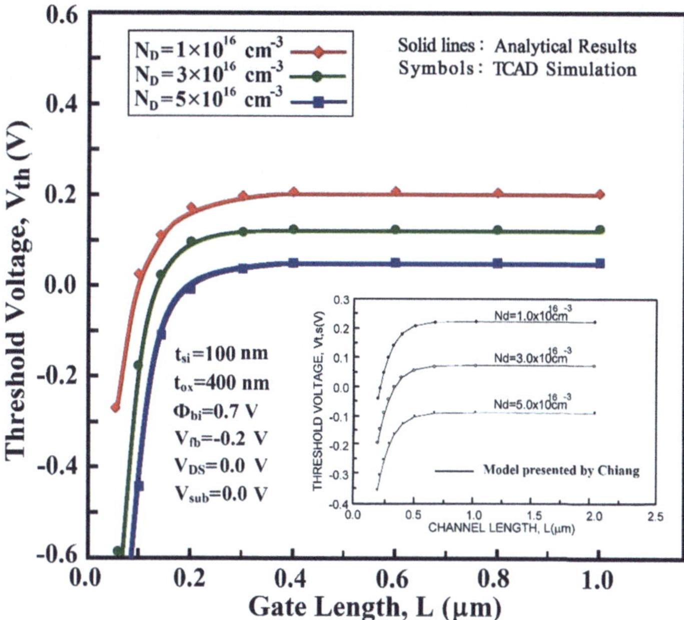

Fig. 6.4 The plot of threshold voltage versus the channel length for various channel doping concentrations. Solid lines correspond to the model, and symbols correspond to the simulation results. (Inset) The analytical results obtained by Chiang

and $t_{ox} = 400$ nm. It exposes that a lower-threshold voltage is achieved as the channel doping increases. For shorter-channel devices, the threshold voltage changes into negative, and the MESFET becomes normally-on. The lower-threshold voltage is favorable when SOI MESFETs are applied to low-power applications. Also, more accordance between our model and TCAD simulation than Chiang's model is seen.

Besides the parameters related to the device structure, the applied biases can provide the control of threshold voltage. Figure 6.5 displays the effect of drain bias on the threshold voltage using three values of drain voltage ($V_{DS} = 0$, $V_{DS} = 1$, and $V_{DS} = 2$ V). It is evident from the figure that, if the gate length is long enough, the variation of threshold voltage with the drain bias is negligible compared to the same shift of threshold voltage for the shorter gate lengths. For instance, the threshold voltage variations for $V_{DS} = 2$ V are several times larger than for $V_{DS} = 0$. Again, the analytical model shows an outstanding agreement with the simulation results for different values of drain–source voltages.

In addition to the drain voltage, the substrate voltage also plays a key role in the threshold voltage. Figure 6.6 illustrates the influence of the substrate bias on the threshold voltage for three values of substrate voltage ($V_{sub} = 0$, $V_{sub} = -1$, and $V_{sub} = -2$ V). As shown in the figure, the negative bias of substrate pulls down the bottom potential and accordingly prevents the formation of the channel. Also, a

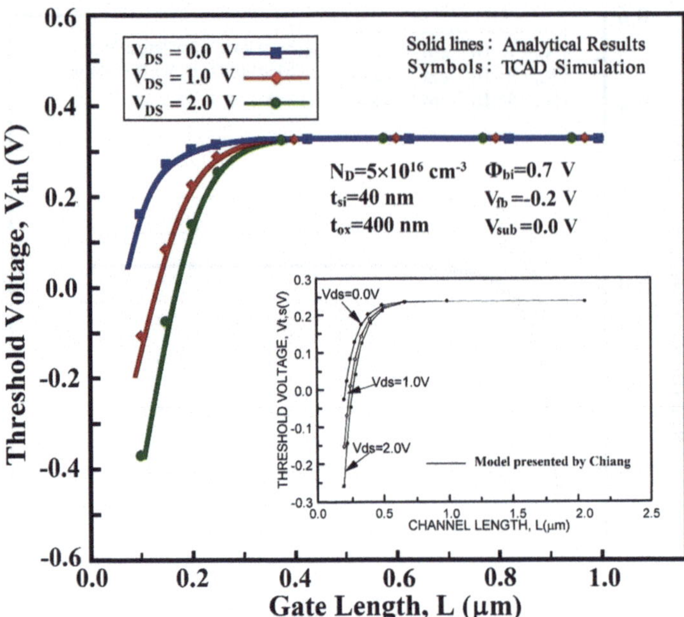

Fig. 6.5 The plot of threshold voltage versus the channel length for various drain biases. Solid lines correspond to the model, and symbols correspond to the simulation results. (Inset) The analytical results obtained by Chiang

further negative of substrate bias can reduce the DIBL effect and considerably improve the degradation of the threshold voltage. Consequently, the substrate bias of SOI MESFET, as well as structure parameters, can be used for modulation of the threshold voltage. This provides more flexibility in the design of SOI MESFET devices. Again, in both Figs. 6.5 and 6.6, our model has a better matching with the simulation results than Chiang's model.

6.2.4 Analysis of DIBL

As seen in Fig. 6.1, the minimum bottom potential is elevated by increasing the drain–source voltage. This effect is called DIBL and originates from the electric field induced by the drain voltage. DIBL increases the drain current and affects the subthreshold performance of the device as the channel length decreases. Figure 6.7 exhibits the change in DIBL with the channel length. In this figure, the DIBL is calculated from the difference value between the threshold voltage with low drain voltage, $V_{DS}(\text{low}) = 0.1$ V, and three values of high drain voltage ($V_{DS}(\text{high}) = 0.5$ V and 1.0 V) as written below in:

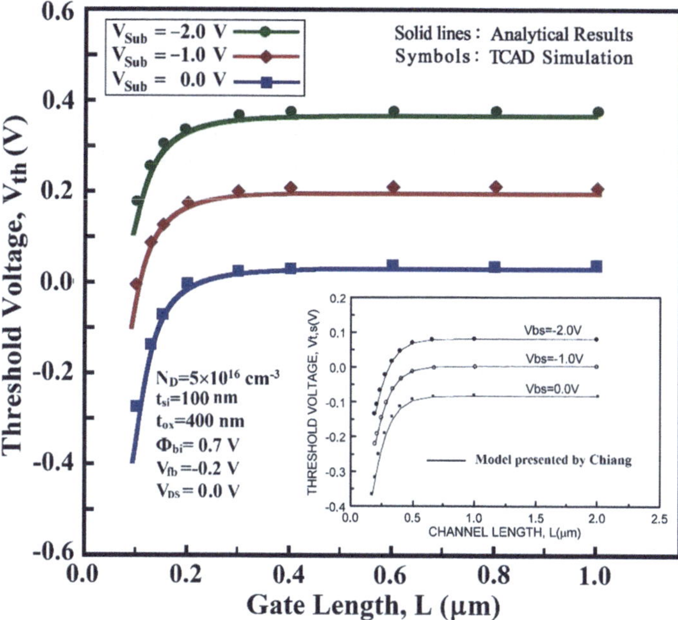

Fig. 6.6 The plot of threshold voltage versus the channel length for various substrate biases. Solid lines correspond to the model, and symbols correspond to the simulation results. (Inset) The analytical results obtained by Chiang

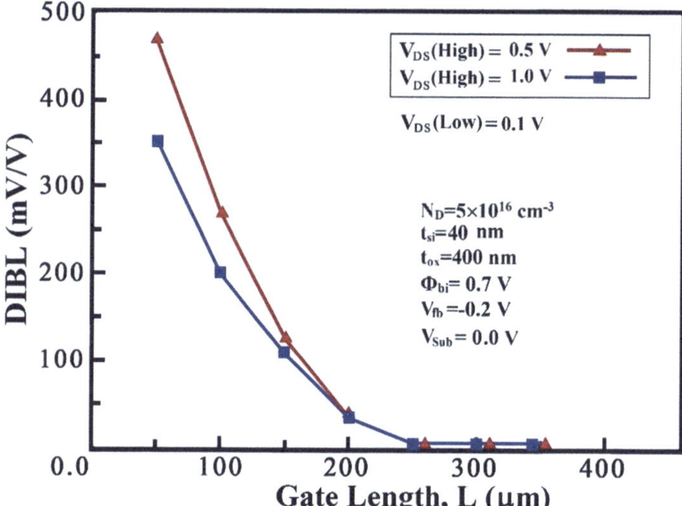

Fig. 6.7 The plot of DIBL against channel length for different drain biases

$$\mathrm{DIBL} = \frac{V_{\mathrm{th}}\big|_{V_{DS(\mathrm{low})}} - V_{\mathrm{th}}\big|_{V_{DS(\mathrm{high})}}}{V_{DS(\mathrm{high})} - V_{DS(\mathrm{low})}} \tag{6.1}$$

It is evident from the figure that for the channel lengths below 100 nm, the DIBL becomes more prominent. The graph indicates that low drain bias can efficiently suppress DIBL.

6.2.5 *Analysis of Subthreshold Current and Subthreshold Swing*

Figure 6.8 shows the computed and simulated results for the variation of subthreshold current against the gate-to-source voltage for three values of drain voltage (V_{DS} = 0.5, V_{DS} = 1.0, and V_{DS} = 2.0 V). The figure indicates increasing of subthreshold current with rising of drain voltage due to the drain-induced barrier lowering (DIBL). The slight change in the slope with the variation of drain voltage represents that the barrier height lowering is weakly reliant on drain voltage.

The last interest is to plot the subthreshold swing. Figure 6.9 displays the subthreshold swing as a function of channel length for various silicon film thickness (t_{si} = 20, t_{si} = 40, and t_{si} = 80 nm) when the thickness of buried oxide is fixed at 300 nm. It is evident from the figure that the subthreshold swing can be alleviated by using thinner silicon film layer. Also in Fig. 6.10, the effect of buried oxide thickness on the subthreshold swing is shown. In this figure, the thickness of silicon film is fixed at 30 nm, and three values of buried oxide thickness (t_{ox} = 200, t_{ox} = 400, and

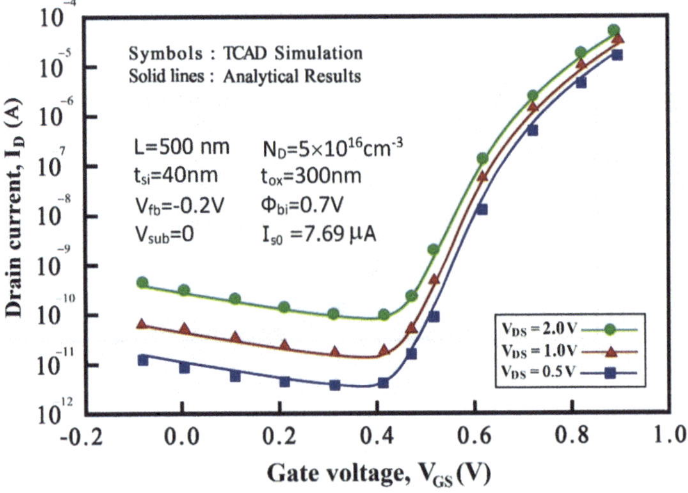

Fig. 6.8 The plot of subthreshold current versus gate voltage for various drain biases

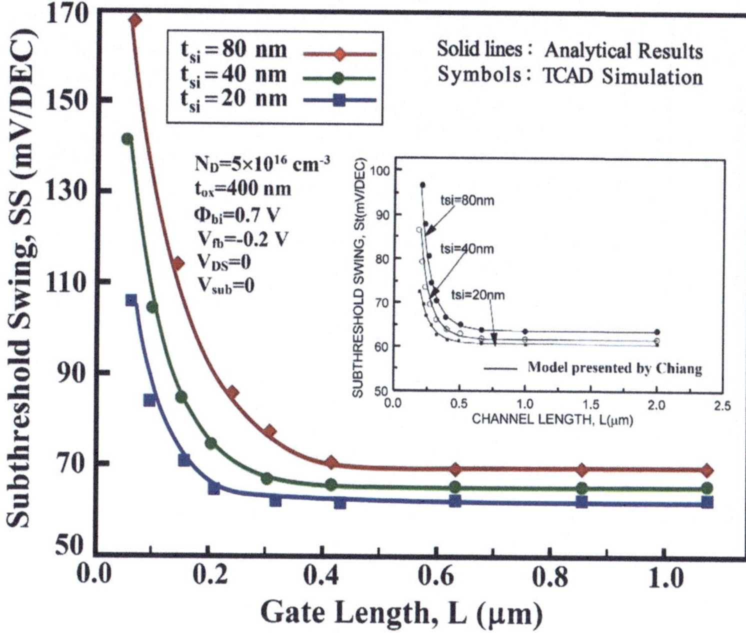

Fig. 6.9 The plot of subthreshold swing versus gate voltage for various silicon film thicknesses. Solid lines correspond to the model, and symbols correspond to simulation data. (Inset) The results obtained by the model presented by Chiang

$t_{ox} = 600$ nm) are used. As seen in the figure, the thicker buried oxide layer decreases the subthreshold swing. Also, our model predicts the subthreshold swing with more accuracy than the model presented by Chiang.

6.3 Analysis of Triple-Material Gate SOI MESFET

In this section, using the analytical model presented in Chap. 4, the performance of triple-material gate SOI MESFET is investigated. Also, to validate the accuracy of the analytical model, the 2-D device simulator ATLAS is used to simulate the different aspects including surface potential, threshold voltage, and subthreshold swing and compare with the results obtained by the analytical model. The structure of an n-channel TMG SOI MESFET is implemented in ATLAS having uniformly doped source, drain, and body regions.

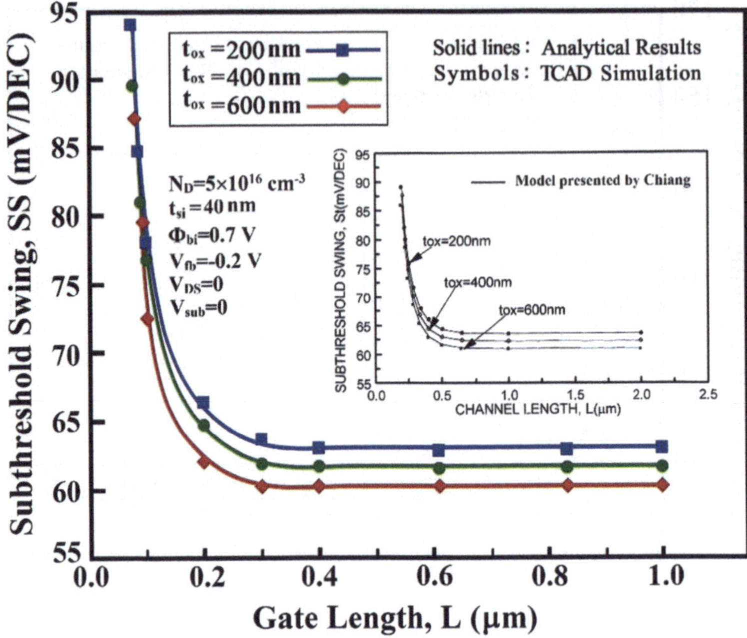

Fig. 6.10 The plot of subthreshold swing versus gate voltage for various thicknesses of buried oxide layer. Solid lines correspond to the model, and symbols correspond to simulation data. (Inset) The results obtained by the model presented by Chiang

6.3.1 Device Parameters and Dimensions

Typical values of the work-function of gate metals used in analysis and simulation are 5.1 eV for M_1, 4.6 eV for M_2, and 4.4 eV for M_3 (see Appendix A). For SMG MESFET, the value of 5.1 eV is chosen for gate metal work-function. All the device parameters of TMG are same as SMG device used in the previous section unless otherwise stated and are given in the text and figure captions. Table 6.2 summarized the dimension and parameters used for investigating triple-material gate SOI MESFET.

6.3.2 Analysis of the Channel Potential

Figure 6.11 displays the simulated and calculated profile of bottom potential of TMG SOI MESFET against the normalized position along the channel for a channel length of $L = 500$ nm and three values of drain voltage ($V_{DS} = 0$, $V_{DS} = 0.5$, and $V_{DS} = 1.0$ V). The simulated data using ATLAS for single-material gate MESFET are also added for comparison. As seen in the figure, there are two step-changes in

Table 6.2 Parameters used for analysis and simulation of TMG SOI MESFET

Specified parameters	Symbol	Value
Work-function of gate metal 1	Φ_{M1}	5.1 eV
Work-function of gate metal 2	Φ_{M2}	4.6 eV
Work-function of gate metal 3	Φ_{M3}	4.4 eV
Channel doping concentration	N_D	5×10^{16} cm^{-3}
Source/Drain regions doping concentration	$N_{S/D}$	10^{20} cm^{-3}
Thickness of back oxide layer	t_{box}	300 nm
Thickness of silicon film	t_{si}	40 nm
Permittivity of silicon	ε_{si}	$11.8 \times 8.85 \times 10^{-14}$ F/cm
Permittivity of SiO_2	ε_{ox}	$3.9 \times 8.85 \times 10^{-14}$ F/cm
Built-in voltage of Schottky junction	Φ_{bi}	0.7 V
Length of gate	L	100–500 nm
The flat-band voltage of SiO_2	V_{fb}	−0.2 V

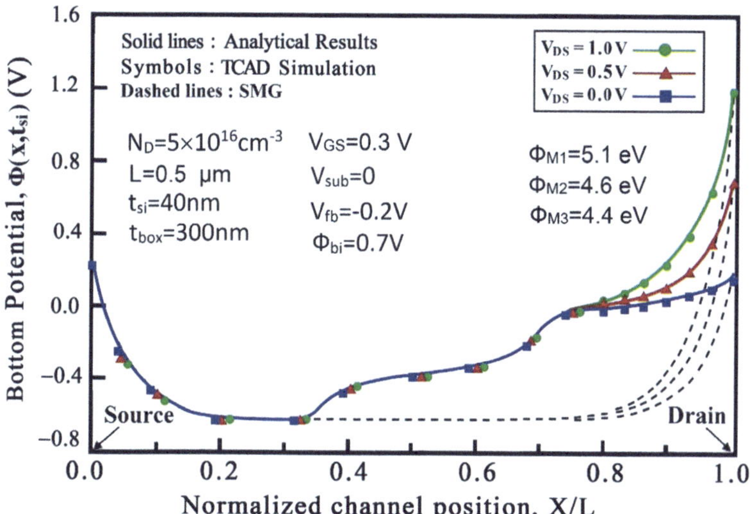

Fig. 6.11 The profile of surface potential for SMG and TMG structures obtained from the analytical model and ATLAS simulation for different drain biases with a channel length $L = 0.5$ μm

the profile of the TMG SOI MESFET which do not exist in the potential of SMG structure. These two step-functions suppress the effect of the electric field induced by the drain voltage in the region under gate metal 1. Due to this effect, there is no significant change in the potential below the metal gate 1, when the drain bias increases. In other words, the channel region under metal 1 is screened from the changes in the drain potential. For this reason, the influence of the drain voltage on drain current after saturation is negligible and consequently, the drain conductance is small. This is the key advantage of the triple-material gate (TMG) structure. Also, regarding Fig. 6.10, the zero-field point for SMG MESFET is located near the drain meanwhile for it lies at the interface of the two metal gates for TMG MESFET. Consequently, DIBL is significantly reduced for TMG structure.

Note that increasing of drain voltage has no considerable change on the minimum bottom potential. This shows that the DIBL is reduced. This reduction of DIBL enhances the subthreshold performance of the device.

Figure 6.12 plots the variation of the bottom potential versus the normalized position along the channel for different ratios of gate materials' length (L_1:L_2:L_3 = 1:2:3, L_1:L_2:L_3 = 1:1:1, and L_1:L_2:L_3 = 3:2:1) keeping the sum of total gate length to be constant. As the result shows, reducing of the gate length M_1 causes the point of minimum potential under M_1 to move toward the source. Accordingly, the peak electric field in the channel shifts more toward the source end, and hence a further uniform electric field in the channel is obtained. As seen, the minimum of bottom potential and its location in the channel are not the same for all three cases. This situation arises because by increasing the L_1:L_2:L_3 ratio, a segment of the channel controlled by the gate metal with higher work-function is also increased. It is necessary to mention that the minimum surface potential between the source side and the minimum channel position can be raised for the high ratio of L_1:L_2:L_3 such as 1:2:3, which consequently enhances the immunity to the SCEs and effectually repressed DIBL for the TMG MESFET. Also, a proper matching between analytical results and simulation data is seen.

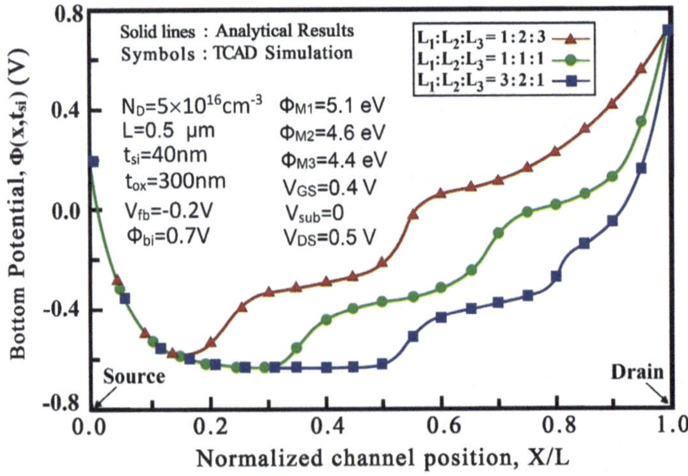

Fig. 6.12 The profile of surface potential for TMG structure obtained from the analytical model and ATLAS simulation for different ratios of L_1:L_2:L_3 with a channel length $L = 0.5$ μm

6.3.3 Analysis of Threshold Voltage

In Fig. 6.13, the profile of threshold voltage roll-off of the proposed TMG MESFET as a function of a channel length at a drain bias of $V_{DS} = 1$ V is plotted and compared with both SMG and DMG SOI MESFETs. As shown in the figure, for more extended channel lengths, the threshold voltage roll-off in all SMG, DMG, and TMG devices remains constant. It is observed that SCEs become serious on decreasing the channel length. However, for the channel length less than 0.2 μm, the TMG device exhibits better subthreshold performance. Also, the offered model is compared with the simulation results where a good accordance is observed. At a constant channel length, the location of the step potentials can be changed for different values of the L_1:L_2:L_3 ratios. In Fig. 6.14, the variation of threshold voltage as a function of the gate work-function ratio for several L_1:L_2:L_3 ratios (L_1:L_2:L_3 = 1:2:3, L_1:L_2:L_3 = 1:1:1, and L_1:L_2:L_3 = 3:2:1) is plotted. From the figure, for certain values of work-function for gate materials, increasing the ratio rises the threshold voltage especially in short-channel devices. This effect leads to a lowering of V_{DIBL} and a consequent decrease in the influence of drain electric field on the channel.

In Fig. 6.15, the dependency of the threshold voltage on channel length for several values of work-functions of metal 1 is demonstrated. As shown in the figure, increasing of work-function of metal 1 elevates the threshold voltage. Meanwhile, the threshold voltage is almost constant for all cases.

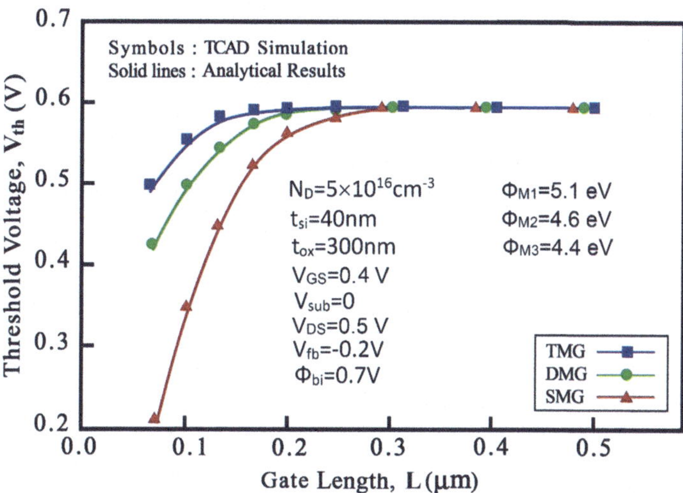

Fig. 6.13 The plot of threshold voltage versus gate length of TMG, DMG, and SMG SOI MESFET

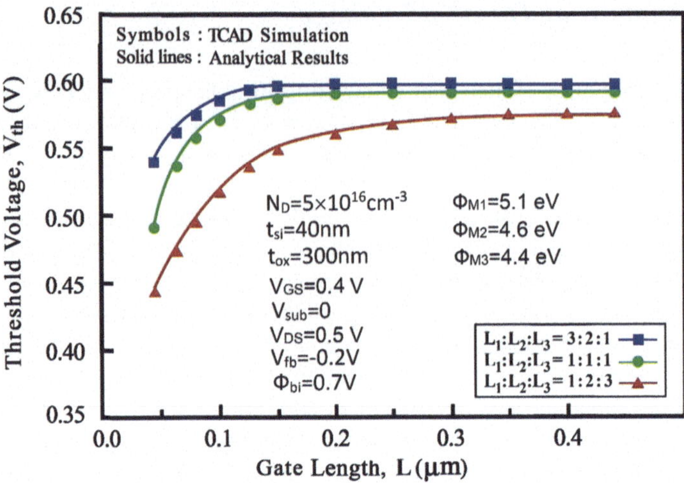

Fig. 6.14 The plot of threshold voltage versus gate length for different ratios of L_1:L_2:L_3

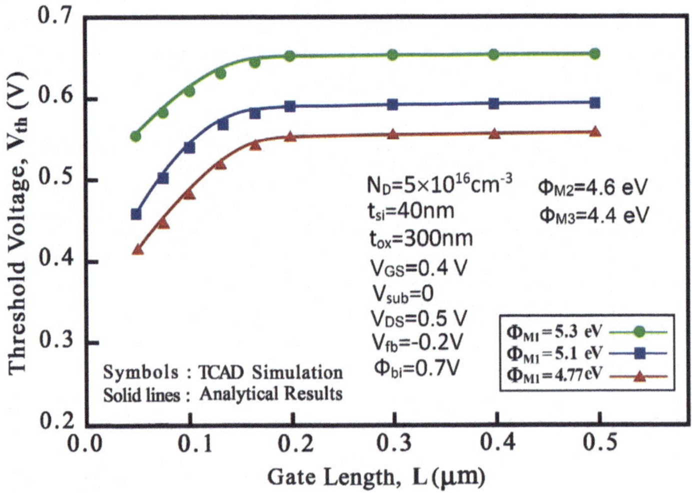

Fig. 6.15 The plot of threshold voltage versus gate length for different gate work-function values of Φ_{M1}

6.3.4 Analysis of Subthreshold Swing

In Fig. 6.16, the dependence of subthreshold swing of TMG SOI MESFET on the channel length is plotted and compared with DMG and SMG MESFET.

It can be seen that the TMG SOI MESFET shows a better subthreshold swing performance when the channel length decreases into the submicron scale. It exposes that the degradation of subthreshold swing in deep submicron scale can be lessened

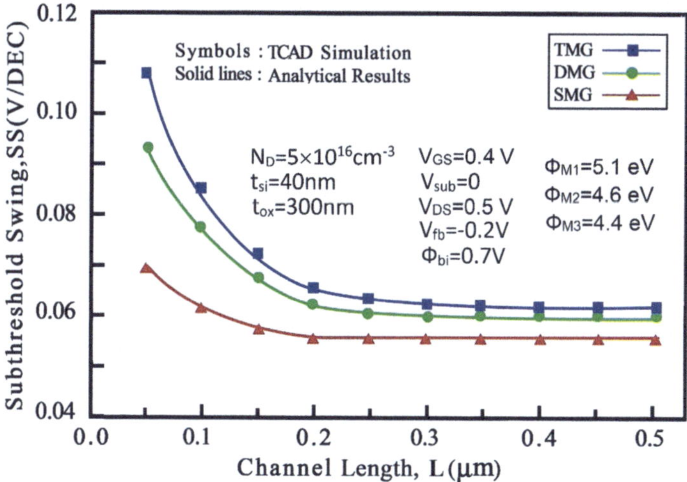

Fig. 6.16 The plot of subthreshold swing versus gate length for SMG, DMG, and TMG MESFET

by the use of the proposed TMG design. Thus, the proposed model offers good immunity against the SCEs and the HCEs.

6.4 Analysis of Three-Gate SOI MESFET

In this section, we have offered some analytical results to demonstrate the subthreshold performance of three-gate SOI MESFET under study considering different device parameters and bias conditions. The accuracy of the analytical model is examined using the TCAD simulator ATLAS from Silvaco [1]. In all figures, solid lines and symbols are used to display the simulation and analytical results, respectively. The revealing values of device parameters and bias condition used for analysis and simulation are listed in Table 6.3.

6.4.1 Analysis of the Channel Potential

The calculated and simulated results of the channel potential against normalized position along the channel for three values of channel length ($L = 50$, $L = 80$, and $L = 100$ nm) are plotted and compared in Fig. 6.17. As seen in the figure, if the channel is long enough, the channel potential in the middle of the channel is nearly constant, but it exponentially increases from the middle of the channel toward the source and drain. If the channel length reduces, due to the SCEs, the minimum potential rises and changes from a broad area to a bottom point of a parabolic curve.

Table 6.3 Parameters used for analysis and simulation of three-gate SOI MESFET

Specified parameters	Symbol	Value
Channel length	L	50–150 nm
Channel width	W	20–60 nm
Channel high	H	20–60 nm
Buried oxide thickness	t_{ox}	400 nm
Channel doping density	N_D	10^{19} cm^{-3}
Flat-band voltage	V_{fb}	−0.2 V
Built-in voltage of Schottky barrier	V_{bi}	0.7 V
Permittivity of silicon	ε_{si}	$11.8 \times 8.85 \times 10^{-14}$ F/cm
Permittivity of oxide	ε_{ox}	$3.9 \times 8.85 \times 10^{-14}$ F/cm
The flat-band voltage of SiO_2	V_{fb}	−0.2 V

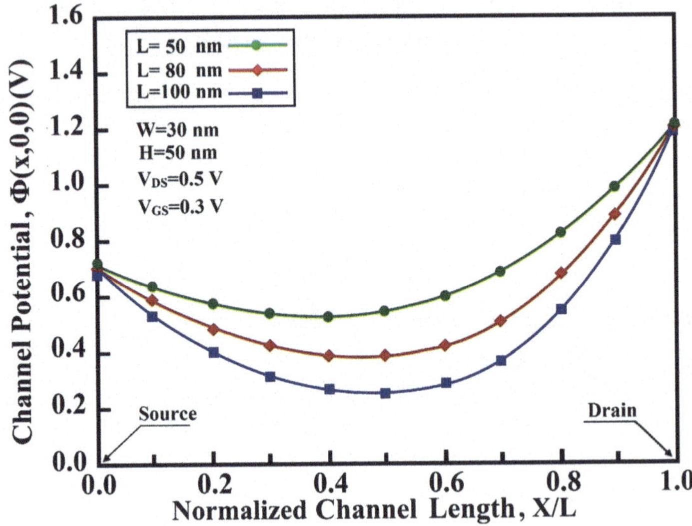

Fig. 6.17 The variation of channel potential against normalized channel length for different values of gate length attained by taking transversal cross-section at $x = L/2$. Solid lines correspond to the model, and symbols correspond to the simulation results

Further, the location of minimum channel potential moves toward the source due to the drain-induced barrier lowering (DIBL). It is depicted that the surface potential obtained from the analytical model agrees well with the results obtained from TCAD simulation.

Figure 6.18 illustrates the variation of the channel potential versus the normalized position along the channel width for three values of channel width ($W = 20$, $W = 40$, and $W = 60$ nm) when the channel length and channel height are fixed at $L = 80$ nm and $H = 50$ nm. From the figure, for all values of channel width, the surface potential at sidewalls of the channel ($z = -$w/2 and $z =$ w/2) is lower than the potential at the middle of channel width (z = 0). Moreover, for the long width device, the surface

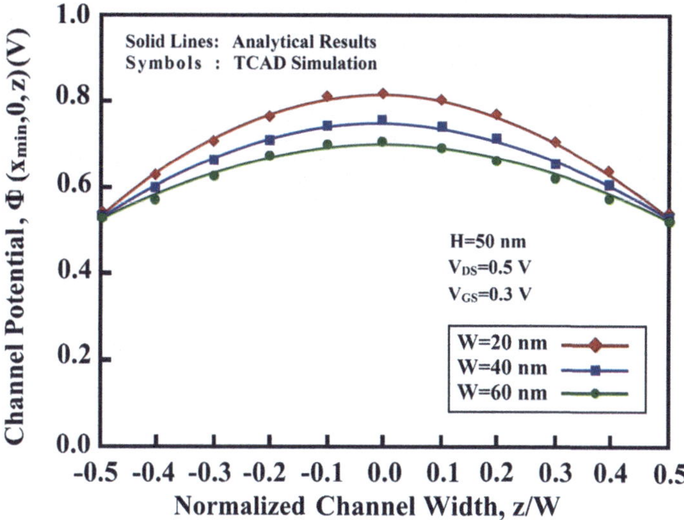

Fig. 6.18 The variation of channel potential against normalized channel width for different values of channel width attained by taking transversal cross-section at $z = 0$. Solid lines correspond to the model, and symbols correspond to the simulation results

potential along the channel width is almost constant, and a broad minimum potential is seen. For the device with shorter width, the potential is elevated and gets to the minimum point in the middle of channel width.

6.4.2 Analysis of the Threshold Voltage

In the following, using the analytical results computed from the model and ATLAS simulation, the threshold voltage of the three-gate SOI MESFET under study has been investigated. The profile of threshold voltage for different values of channel length, channel width, and drain voltage are plotted. It is noticed that the geometry limitation of $W_{\text{fin}} < H_{\text{fin}}$ is required for achieving realistic and operational FinFETs of channel length $L < 100$ nm [2]. Figure 6.19 shows the threshold voltage versus channel length for two values of channel width ($W = 20$ nm and $W = 40$ nm). It is evident that if the channel is long enough ($L > 80$ nm) for both chosen fin widths, the threshold voltage is nearly constant at about 0.75 V. If the channel length declines, the threshold voltage exponentially decreases owing to increase of SCEs.

Figure 6.20 shows the variation of threshold voltage against the channel width extracted from analytical model and simulation data for three values of gate length (L = 30 nm, L = 50 nm, and L = 100 nm). From the figure, while channel length and channel height are constant and the fin width decreases, the gate has more control on channel and threshold voltage increases. Also, the good match between two groups

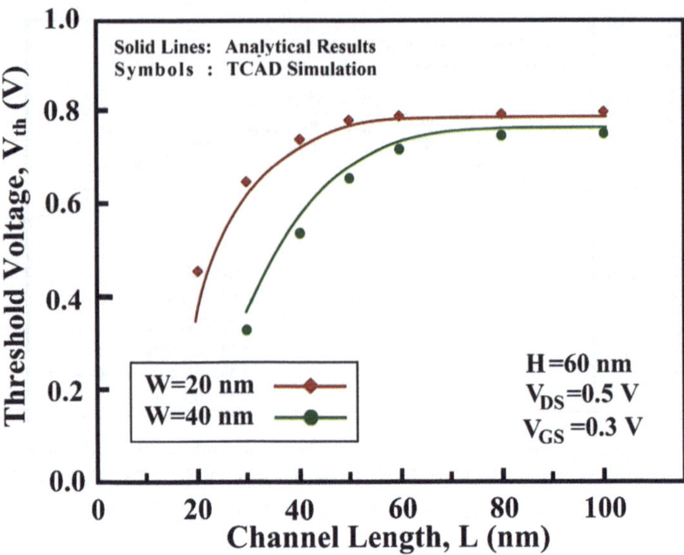

Fig. 6.19 The variation of threshold voltage versus gate length for two values of channel width. Solid lines correspond to the model, and symbols correspond to the simulation results

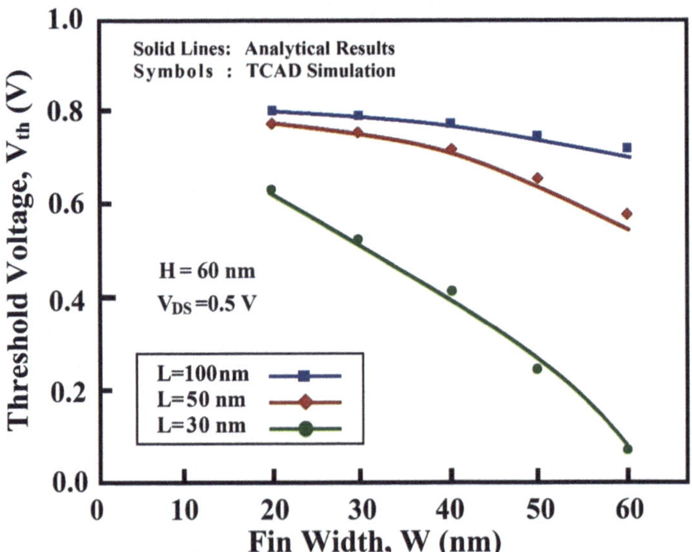

Fig. 6.20 The variation of threshold voltage versus fin width for different values of channel length. Solid lines correspond to the model, and symbols correspond to the simulation results

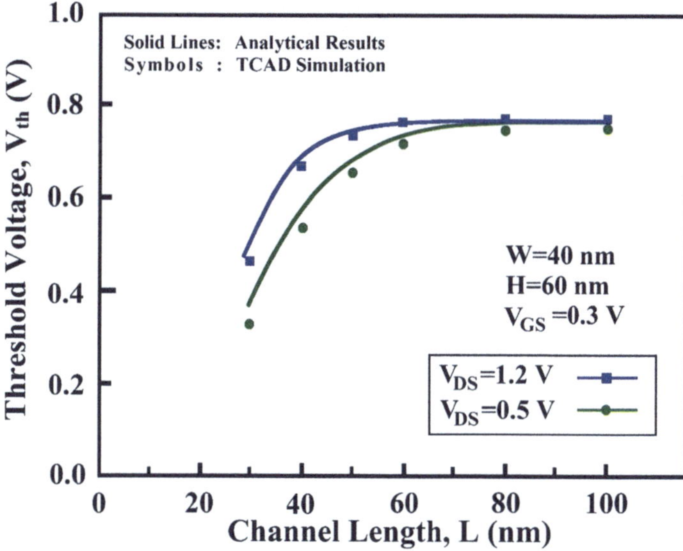

Fig. 6.21 The variation of threshold voltage versus channel length for different values of drain voltage. Solid lines correspond to the model, and symbols correspond to the simulation results

of results is exposed. The threshold voltage is further verified, in Fig. 6.21, by studying the effect of drain voltage. As seen in the figure, if the transistor channel is sufficiently long ($L > 80$ nm), the threshold voltage is constant and independent of the drain voltage. For shorter-channel devices, the threshold voltage decreases with increasing of drain voltage. It can be seen that the analytical model can properly predict the threshold voltage roll-off for various channel dimensions and drain biases.

References

1. S.K. Saha, Technology Computer Aided Design, CRC Press, 17–60 (2016)
2. N. Fasarakis, A. Tsormpatzoglou, D.H. Tassis, C.A. Dimitriadis, K. Papathanasiou, J. Jomaah, G. Ghibaudo, Analytical unified threshold voltage model of short-channel FinFETs and implementation. Solid State Electron. **64**, 34–41 (2011)

Chapter 7
Future Works on Silicon-on-Insulator Metal–Semiconductor Field Effect Transistors (SOI MESFETs)

7.1 Conclusion

Our study was motivated by the principle that introducing new designs for SOI MESFETs can take advantage of both non-classical structure and SOI MESFET while avoiding the disadvantages related to the conventional MOS devices. Accordingly, we have presented two new high-performance non-classical designs: the triple-material gate silicon-on-insulator metal–semiconductor field effect transistor (TMG SOI MESFET) and the three-gate silicon-on-insulator metal–semiconductor field effect transistor (TG SOI MESFET) for the first time and then investigated them extensively through analytical and simulation methods. We studied the most critical issues including channel potential, threshold voltage, subthreshold swing, and short-channel effects that are the primary concern in recent years due to the continuous scaling of FETs into the deep-submicron territory. Moreover, we have demonstrated the superior subthreshold performance of the proposed MESFET designs by comparison with classical SOI MESFET. According to this, we introduced and discussed the previous analytical models presented for classical SOI MESFET developed in recent years and then we offered a new exact analytical model to display the subthreshold behavior of the device. In summary, the conclusions resulting from this work are divided into three parts:

(a) Developing an accurate analytical model for fully depleted (FD) short-channel SOI MOSFET which can precisely describe the performance of the device.
(b) Introducing, modeling, and simulation of tri-material gate (TMG) SOI MESFET.
(c) Introducing, modeling, and simulation of three-gate (TG) SOI MESFET.

I. S. Amiri et al., *Device Physics, Modeling, Technology, and Analysis for Silicon MESFET*, https://doi.org/10.1007/978-3-030-04513-5_7

7.1.1 *Fully Depleted Short-Channel SOI MESFET*

The analytical model of potential distribution along the channel in an FD-SOI MESFET device is obtained by solving the 2-D Poisson's equation using a superposition method along with the separation of variables. Based on the surface potential model, analytical expressions for threshold voltage and subthreshold swing are derived. The conclusions are:

- The model can display the subthreshold behavior of the device including potential distribution along the channel, threshold voltage, short-channel effects, and subthreshold swing for different channel lengths.
- The model can predict the outcome of various device parameters like body doping concentration, silicon film thickness, and buried oxide thickness.
- The model can predict the impact of applied drain and substrate biases.

7.1.2 *New Design Tri-Material Gate (TMG) SOI MESFET*

To combine the advantages of silicon SOI MESFET and gate engineering, a new device, tri-material gate (TMG) SOI MESFET, has been offered and expected to diminish the short-channel effects (SCE's) and enhance carrier transport in the active channel. An analytical model of potential distribution along the channel is obtained by solving the 2-D Poisson's equation using a superposition method along with the separation of variables. Then the surface potential model is used to derive analytical expressions for threshold voltage and subthreshold swing. Using the obtained model, and also TCAD simulation, the subthreshold performance of the device are studied, plotted, and compared with classical SOI MESFET. It is concluded that:

- The presented analytical model predicts the presence of two step-functions in the profile of channel potential due to the three different work-functions of three gate materials.
- Triple-material gate SOI MESFET shows better submicron performance than classical SOI MESFET.
- The potential distribution along the channel depends on the values of gate material work-functions and also the length of the three gates. With the proper combination of work-functions and the gate materials, better electrical performance can be obtained.
- The short-channel effects like threshold voltage roll off, and DIBL is lesser for TMG structure compared to SMG and DMG structures.

7.1.3 New Design Three-Gate (TG) SOI MESFET

In order to take advantage of the properties of the SOI MOSFET and FinFET in a single transistor, we propose SOI three-gate MESFET. Also, a 3-D analytic model which can exactly describe the performance of the recommended device is developed. Using the gained model and TCAD simulation, the subthreshold performance of the device is studied, plotted, and compared with classical SOI MESFET. It is concluded that:

- The TG SOI MESFET addresses the concerns of scaled CMOS at extremely short-channel lengths, by offering the short-channel control of the three-gate architecture.
- The TG SOI MESFET is looked upon as an upcoming candidate for next generation of transistors.

7.2 Future Scope of the Work

This research may have some possible extensions that could be considered as continuing research work. Some special recommendations regarding this work are as follows:

- The proposed non-classical architecture presented in this study can be well extended to partially depleted SOI MESFETs.
- In this research, the idea of TMG is used in the front gate of the MESFET. Another innovative structure may be proposed if that idea is used in back gate also (double-gate TMG MESFET).
- The idea of TMG may be offered for TG MESFET.
- Experimental results can be used for more confirmation of the efficiency of the proposed structures.
- The presented devices can be investigated at the circuit level, e.g., inverter. Then, the performance of the circuit composed of the proposed devices can be compared with a similar circuit that is composed of the classical transistors.

Appendix: The Work Functions of the Common Metals

Metal	Symbol	Work function (eV)	Metal	Symbol	Work function (eV)
Aluminum	Al	4.08	Ruthenium	Ru	4.71
Tantalum	Ta	4.19	Rhodium	Rh	4.8
Molybdenum	Mo	4.2	Cobalt	Co	4.97
Strontium	Sr	4.21	Palladium	Pd	5
Vanadium	V	4.3	Nickel	Ni	5.1
Titanium	Ti	4.33	Rhenium	Re	5.1
Stannum	Sn	4.42	Aurum	Au	5.1
Tungsten	W	4.52	Iridium	Ir	5.27
Chromium	Cr	4.6	Platinum	Pt	5.34

I. S. Amiri et al., *Device Physics, Modeling, Technology, and Analysis for Silicon MESFET*, https://doi.org/10.1007/978-3-030-04513-5

Index

I. S. Amiri et al., *Device Physics, Modeling, Technology, and Analysis for Silicon MESFET*, https://doi.org/10.1007/978-3-030-04513-5

P

Q

S

T

The manufacturer's authorised representative in the EU is Springer Nature Customer Service Centre GmbH, Europaplatz 3, 69115 Heidelberg, Germany. If you have any concerns regarding our products, please contact ProductSafety@springernature.com

Printed and bound by CPI Group (UK) Ltd, Croydon, CR0 4YY
18/07/2026
02171007-0001